DISCOVERING THE MICROSCOPIC WORLD

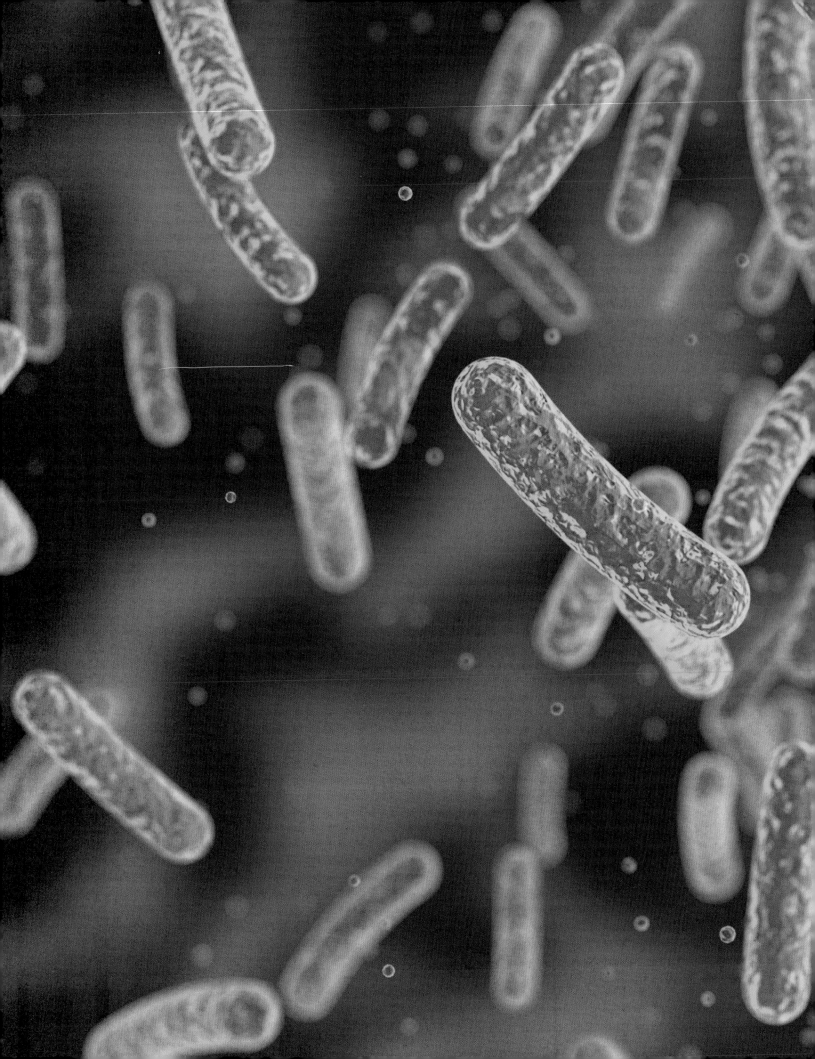

DISCOVERING THE MICROSCOPIC WORLD

A guide to the incredible structures of organisms

Marianne Taylor

SIRIUS

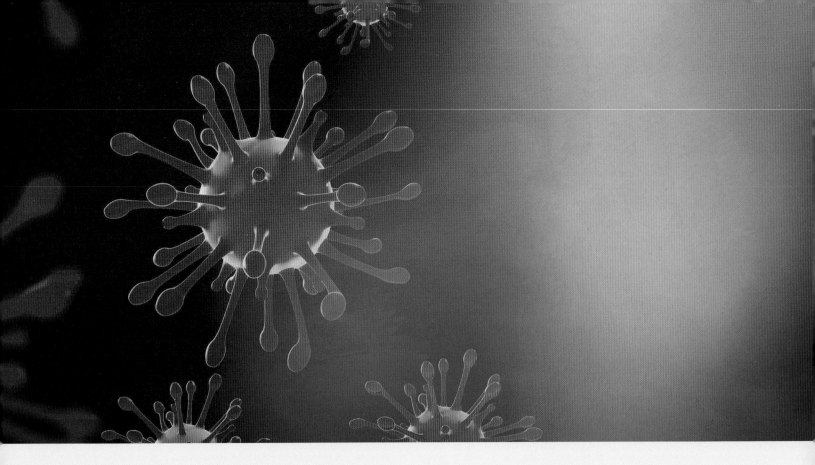

SIRIUS

This edition published in 2023 by Sirius Publishing, a division of
Arcturus Publishing Limited,
26/27 Bickels Yard, 151–153 Bermondsey Street,
London SE1 3HA

ISBN: 978-1-3988-3052-3
AD010550US

Printed in China

Contents

// Introduction

Above *The world's smallest insect is a fairyfly (family Myrmaridae) that is less than 0.005 in (0.14 mm) long.*

Did you know that the vast majority of cells in your body are not human? Did you ever wonder how complex life forms evolved from biochemical beginnings, or what a virus wants to get out of its life (if it can even be considered alive)? Today, we can see and marvel at a glorious array of animals, plants, and fungi on Earth, including the largest animal ever to have lived, but our earliest scientists were aware (or at least suspected) that there was also a world of life too small for us to see with our own eyes.

Microscopic life, whether individual living things or the tiny building blocks that make up larger bodies, was mysterious to the point of magical to us for millennia. We knew there were living animals that were on the very threshold of the size our eyes could discern, so why not even smaller? After all, we knew how animal life cycles worked, and an insect 0.04 in (1 mm) long, resembling a moving speck of dust to our eyes, must have grown from an even tinier larva, which hatched from a still smaller egg. We also observed that there were life processes that could not be fully explained by what we could see, such as disease spread, the healing of wounds, the conception of life, and the breakdown of organic matter— but that didn't mean we needed to invoke the supernatural, just the very, very small. Through centuries of innovation, we have found ways to infer and understand the workings of microscopic life, and (through increasingly clever and sophisticated optical technology) we can literally see into this world, in ever increasing detail.

Simple beginnings

Biology is concerned with life and its processes, but as we dive deeper and deeper into the microscopic world, we move from biology into chemistry, and from chemistry to physics. We may never find a way to be able to see the fundamental particles from which atoms are built, but physicists have learned and continue to learn much about their nature, and the often baffling ways in which they behave. When atoms bond together to form compounds, such as the two hydrogen atoms and one oxygen atom that form a single molecule of the compound we call "water," we are in the realm of chemistry, which describes the elements from which all matter is formed, and how compounds form and interact.

Organic chemistry is concerned with a particular group of chemical compounds—those that contain the element carbon. Most also contain hydrogen, and often also oxygen and nitrogen. These compounds are of particular interest to us because they are the components of living things—their tissues and bodies. Organic compounds include sugars and proteins, fatty acids, RNA, and DNA—everything you need to form a complete, functional organism. The simplest free-living entity that meets some of the criteria for being "alive" is a viroid, which is nothing more than a loop-shaped RNA molecule. Biochemistry brings the two major sciences of chemistry and biology together, as it concerns the chemical processes that relate to living organisms and tissues. This field of study is central to our understanding of the microscopic world. However, many other aspects of biology will come into play too as we explore the microscopic world, including ecology, evolution, anatomy, and physiology, and behavioral biology.

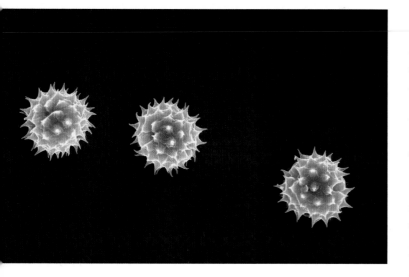

Above *The scanning electron microscope reveals the beautiful detail of these daisy pollen grains.*

Beauty in miniature

If you peer at a small object through a convex glass lens, your view of it will be magnified. The hand lens is a great tool for an entomologist, botanist, or mycologist working in the field, who would like a more detailed view of insects, plant parts, and fungi respectively. A basic microscope uses the same lens technology but is designed to view tinier objects, so requires more lenses and also a way to hold the sample item very steady, because any jiggling movement is magnified too. The sample is typically placed on a glass slide, often with a very fine glass cover, and the slide slots into place. The microscope usually also has a light source to illuminate the sample. You may have used this piece of equipment in school to examine a few drops of pond water, teeming with tiny animal life, or perhaps to look at the cells in a slice of liver or leaf, or even a swab from the inside of your own cheek.

For looking at objects in more detail and under higher magnifications, you will require a scanning electron microscope. This uses an accelerated electron beam as its light source, rather than conventional lighting. The much shorter wavelength of electrons compared to ordinary light photons means that this kind of microscope can resolve far more detail. Such microscopes are usually combined with specialized cameras to capture images (electron micrographs). Through this technology we can visualize and take useful images of cells and smaller objects in beautiful detail.

Above *Bacteria of the genus* Campylobacter *have a spiral structure.*

About this book

Chapter 1 of this book begins with the first chemical processes associated with life, and from there explores the simplest "life-like" entities, and how they do and do not meet the various criteria that we consider necessary to qualify them as an actual living organism. Viruses are included in this category, and we explore their structure, biological processes, and diversity. Subsequent chapters look at unambiguously living things—the simple prokaryotes, which include bacteria and constitute two of life's three domains, making up the vast majority of life on Earth, and at the more complex eukaryotes, from single-celled organisms up to the tiniest members of the familiar animal, plant, and fungus kingdoms (and a few representatives of less familiar groups as well). We also look at the microbiology of the various distinct systems, from roots to flowers, and muscles to brains, of macroscopic organisms—including ourselves.

Left *Feather coloration, as seen here in normal and diluted forms in two gentoo penguins, is produced by deposition of pigment molecules in the feather's microstructure.*

RNA, DNA, AND VIRUSES

All matter is made up of chemical elements. These elements may exist in their "pure" form— for example, a nugget of gold. More often, we find them in combination with other elements to form molecular compounds—a familiar example is sea salt, a compound of the elements sodium and chlorine. The compounds that we find in living organisms are primarily made up of carbon, hydrogen, oxygen, and nitrogen. Some such organic compounds have the ability to self-replicate, and these molecules—RNA and DNA—are the foundation of life on Earth. They can be found in all living things, but also exist in "near-life" entities such as viroids and viruses, from which our world's truly living things are likely to have evolved.

The iconic double-helix coils of DNA hold the answers to so many questions we may ask about our species, our evolution, and even our own individual traits.

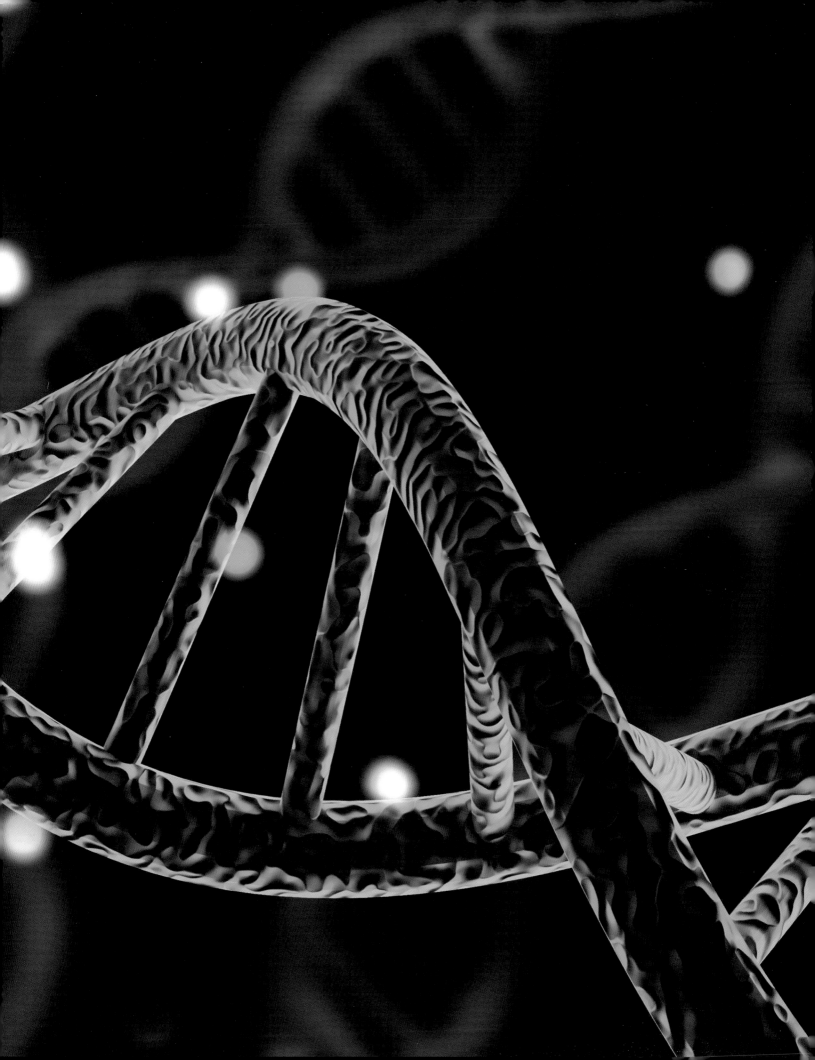

// The chemistry of life

Our planet was formed from a cloud of dust and gas that spun out from our forming sun more than 4.5 billion years ago. The other planets in our solar system were also formed out of sun debris. The Sun is made mostly of hydrogen, the lightest element, and (because of the huge gravitational pressure it is under) this hydrogen is not a gas but a very energetically active plasma form. The material that spun far enough from the gravitational pull of the forming Sun collected into several objects that orbit the sun and are heavy enough to have their own strong gravity. These are the planets of our solar system.

Those closest to the sun (Mercury, Venus, Earth, and Mars) are rocky and made of heavier elements and compounds, because the Sun's heat gives hydrogen and helium too much energy for the planets' gravity to hang on to them. Instead their atmospheres are made of heavier gases. Further out are the much bigger gas giants (Jupiter, Saturn, Uranus, and Neptune), which are far enough from the Sun to hold on to a thick atmosphere of hydrogen and helium,

Below *Lightning strikes and other high-energy events, acting on water and atmospheric gases, could have created the first organic molecules.*

Above *All of the planets in our solar system were formed from matter cast off from the young Sun.*

around a solid core of heavier elements. These gassy planets are themselves orbited by rocky moons (lots of them, in the case of Jupiter and Saturn). Throughout the solar system, Earth (by virtue of its distance from the Sun) is the only body with abundant surface water in liquid form, and this is the key to it being able to support life.

In its very early days, the Earth was a ball of molten rock with a carbon dioxide atmosphere, but it cooled down quickly and its rocky crust formed, with liquid water on top. The cooling atmosphere now held water vapor and nitrogen (in the form of stable two-atom molecules—N_2) as well as carbon dioxide. Nitrogen is one of the four elements most vital to life, being a component of proteins, and between them, the water and the carbon dioxide provided the other three—carbon, oxygen, and hydrogen.

ABIOGENESIS

The idea of life forming spontaneously from non-living materials is called abiogenesis, and it is a tricky concept to get behind. We all know that if you want to make new life you need existing life, and we don't even have a way of restoring dead organisms to life (or any reason to believe that it could ever be possible). But early Earth's conditions could not have supported life so, unless you believe living things arrived on Earth from elsewhere, then abiogenesis must have occurred. We know that the elemental building blocks of biochemistry existed on Earth from very early in its history, and we also know that this planet was highly volatile, with storms and solar activity providing the intense energy needed to break apart N_2, H_2O, and CO_2 molecules, as well as other inorganic molecules present in the Earth's seas, and allow them to re-form in different configurations—for example, methane gas (CH_4), the simplest hydrocarbon. Experiments with applying powerful jolts of energy to this mixture of molecules show that a range of simple organic compounds can be produced this way, including components of the nucleic acids that make up RNA.

// Nucleic acids

Most of the cells in your body contain a nucleus, inside which is your DNA, which is stored in paired sets called chromosomes. DNA is a code for building the proteins your body needs. Deoxyribonucleic acid, to give DNA its full name, is a long organic compound, arranged as a double strand with a spiral shape, like a twisted ladder. It is made of small components called nucleic acids, which come in four different chemical types. Put a particular string of different nucleic acids together and you have a gene, which is a complete instruction for building one of the proteins in your body. Each chromosome is formed from hundreds or thousands of genes. The entirety of genes that exist in a species are called its genome. When you were conceived, your genes came from a randomized 50 percent of your mother's DNA and 50 percent of your father's. The differences between your genes and another person's determine many of the differences between your body and theirs.

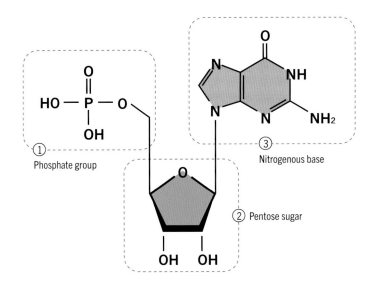

Above *Chemical structure of guanine, one of the five nucleotides that make up DNA and RNA.*

Below *Children inherit a random 50 percent of the genes from each parent, giving a unique combination of traits.*

Differences between DNA & RNA

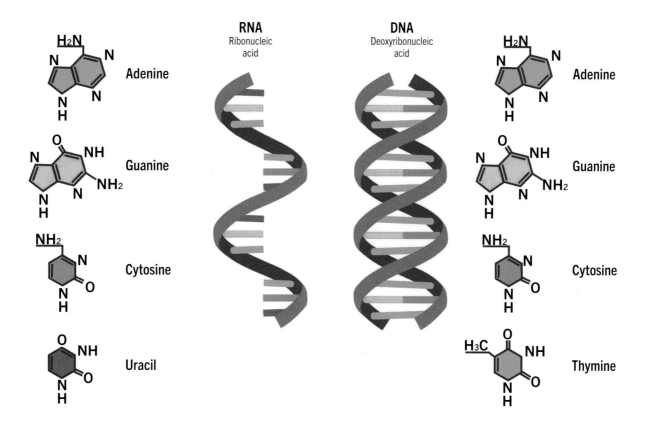

Above *The nucleotides that make up RNA and DNA, and how they pair together in the double-stranded helix of DNA.*

DNA is a nucleic acid and is one of two important nucleic acids that you'll find inside most living eukaryotic cells. The other is RNA (ribonucleic acid), which is also spiraled but usually has just a single strand. In your cells, RNA and DNA work together, with RNA taking part in the process of making proteins. Because they can copy themselves, DNA and the simpler RNA are known as self-replicating molecules. Mistakes occur now and then during the process of copying, so the two copies may have a slightly different sequence of nucleic acids than the original.

The ability to self-replicate, and the way this naturally creates genetic variability on which natural selection could act, is the reason why many biologists believe that the precursor to true life was a population of "free-living" RNA molecules in the early Earth's waters. For example, an RNA molecule may have a genetic difference that makes it self-replicate more quickly, or makes it more stable, than others. If so, it is likely to become more abundant than molecules that replicate more slowly or are less stable. So, over time,

the molecules in this "RNA world" would become better survivors. It is also known that some RNA molecules can behave like enzymes—proteins that facilitate (catalyze) chemical reactions—so they could have helped other RNA molecules to self-replicate.

THE MAKE-UP OF NUCLEIC ACIDS

DNA is built from molecules called nucleotides. Each nucleotide has three components: a sugar and a phosphate group, which form the "spines" of each of the two strands; and a nitrogenous base, each of which bonds with another on the opposite strand, forming the "rungs of the ladder." There are four kinds of nitrogenous bases in DNA. They are adenine and thymine, which always pair up with each other, one on each side of the double strand, and cytosine and guanine, which also pair up. In RNA, thymine is replaced with uracil.

// Self-replication

Ever since COVID-19 became a global emergency, many of us have learned a lot more about how viruses reproduce themselves. They, and the even simpler viroids, need to use host cells to complete the self-replication process but it is likely that their ancestors were some form of free-living entities, able to self-replicate without making use of a host. Larger organisms, such as bacteria and the much more complex single-celled eukaryotes, replicate themselves on their own, while many complex multicellular life forms have evolved sexual reproduction and self-replicate by combining their genes with a partner. At the molecular level, though, the process is all about self-replicating RNA and (in more complex life) DNA. The ability to self-replicate is one of the fundamental qualities we use to determine whether an entity on the threshold of life is truly "alive."

When we examine the nucleus of a living eukaryotic cell under a high-powered microscope, we won't be able to see the sprawling mass of DNA until the cell is ready to divide. At this point, the individual chromosomes separate and their material becomes much more condensed, and

they become visible in the familiar X-shaped form. The processes that happen next, though, are on a scale too small for us to observe. The DNA ladders are "unzipped" and a new matching strand for each half is built (with the help of enzymes), using free-floating nucleotides within the nucleus. Once a complete duplicate set of chromosomes in the nucleus has been built, the nucleus of the cell then divides, with one full set of chromosomes in each half (and then the rest of the cell divides too). The replication process is very fast and very accurate, but with about 3.2 billion nucleotides in the entire genome, it is not surprising that a few copying errors will occur.

RNA replication also occurs in cells, but does not require any unzipping, as the original molecules are already just a single strand. It otherwise works in a similar way, but occurs within the cell cytoplasm, rather than within a cell nucleus. Most viruses have RNA only, as do viroids, and they do not replicate it within themselves but instead co-opt enzymes and structures within their host cells to replicate themselves (we look at this process later in this chapter).

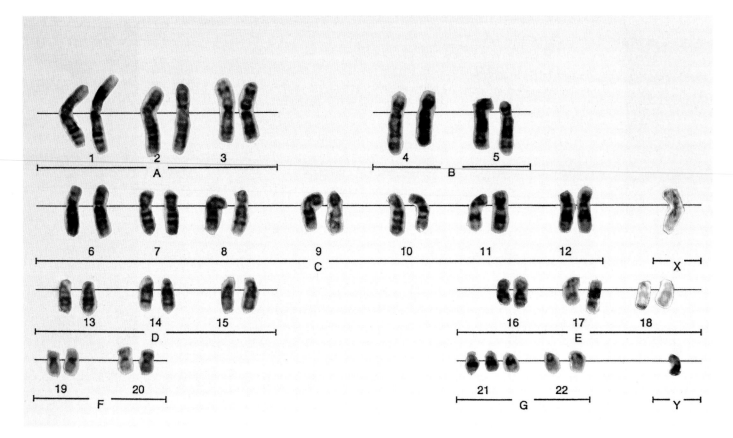

Above *A human karyotype. This individual is male, having one X and one Y chromosome, and also has an extra copy of chromosome 21, a trisomy condition that gives rise to Down's Syndrome.*

DIPLOID AND HAPLOID

A cell that contains pairs of chromosomes is called "diploid." Nearly all the cells in your body are diploid. However, for us and other organisms that reproduce sexually, we also need to make haploid cells—in our case, eggs and sperm, the sex cells or gametes. Each egg and each sperm cell is haploid—it contains only one set of chromosomes, so that when it pairs up with its corresponding gamete, their chromosomes are combined together to form two complete sets. So when the precursor cells to gametes divide, their DNA is not copied.

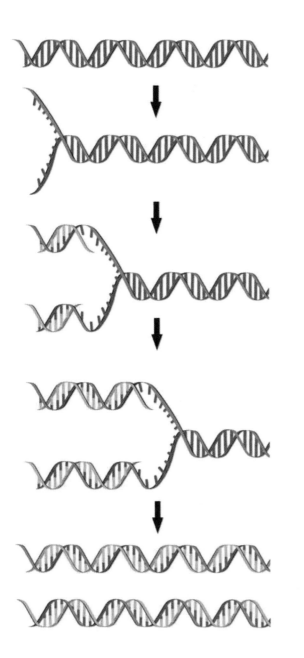

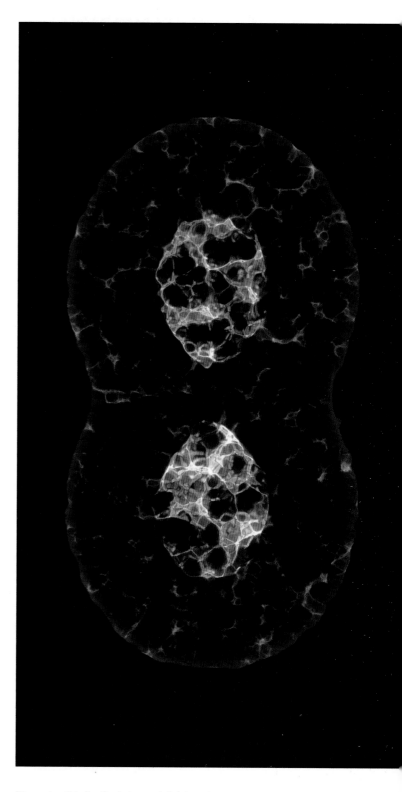

Above *A cell in the final stages of division, the genetic material having been copied and two daughter nuclei formed.*

Left *The process by which DNA is copied, with the "unzipping" of the double helix followed by creation of a duplicate of each strand.*

// Viroids

The tiniest and simplest entities that show some signs of being alive are viroids. An individual viroid is just a loop of RNA, with no enclosing wall or membrane around it. All known viroids live within plants, and usually cause obvious diseases. They replicate themselves in the nuclei of plant cells, using the plant's own enzymes to make this process happen. Life's precursors may have been populations of free-living (and freely self-replicating) RNA molecules, and some biologists think that viroids are modern-day descendants of these "semi-living" inhabitants of that ancient RNA World that would have existed more than 4 billion years ago.

We might have continued to overlook viroids indefinitely if it weren't for the fact that some of them affect commercially important plant species. Discovering the agents that caused various crop-spoiling diseases in potatoes, tomatoes, apples, pears, peaches, and avocados began in 1971, when a Swiss-American plant pathologist, Theodor Otto Diener, isolated the molecular code for the viroid that causes the potato spindle tuber disease. The nature of viroids was further clarified when the agent that causes the disease nicknamed "scalybutt" in citrus fruits was similarly decoded. Today, several families of viroids are recognized, all of them affecting various species of flowering plants.

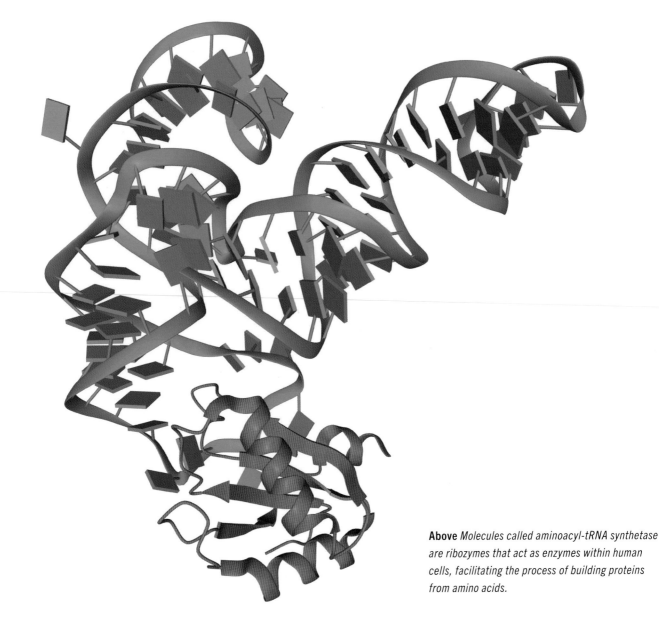

Above *Molecules called aminoacyl-tRNA synthetase are ribozymes that act as enzymes within human cells, facilitating the process of building proteins from amino acids.*

Above *Structure of the RNA molecule that makes up a viroid.*

A viroid consists of a ring-shaped molecule of RNA, and replicates itself using the enzyme RNA polymerase II, found in the cells of its host (and used by the host to replicate its own RNA). It can spread to other plants directly when leaves are in contact, or via insects that feed on the plant tissues. Some viroids don't harm their hosts in any significant way but others cause serious disease.

RIBOZYMES

The idea that viroids descend from free-living RNA molecules is somewhat undermined by the fact that none has ever been found independently of a host plant, and that they cannot replicate without their host. Flowering plants, the only known hosts of viroids, did not evolve until a mere 275 million years ago—so what were viroids (or their ancestors) doing in the intervening billions of years? The discovery of genetic elements called ribozymes has led us to consider an alternative explanation for where viroids came from. Ribozymes are small RNA molecules, very similar to viroids in their make-up, and occur in all kinds of plants and animals, including humans. The genetic code to build ribozymes is found in the part of our DNA that does not code for making proteins (the so-called "junk DNA"). In our cells, ribozymes act as enzymes and have many vital functions, but viroids could be "rogue ribozymes" that have become agents of infection.

Below *A stunted potato plant, affected by a viroid disease.*

// What is the simplest form of life?

Viroids are not generally considered to be alive, but they can and do reproduce themselves, which is one of the traits we have traditionally used to distinguish living things from non-living entities. It might seem trivially easy to tell living things from non-living things when they are big enough for us to see them properly. It's also usually easy for us to tell when an animal has died. Animals in deep hibernation don't move and may not be responsive, but they do still breathe, and their hearts continue to beat. What about a plant, though? It doesn't move at all, nor does it breathe or have a heartbeat. It does grow and does reproduce itself, and it does consume nutrients and produce waste. However, perennial plants die back and regrow, and a still-living tree can have dead branches.

So we know that, once we bring plants into the picture, some of the things that we feel define us and other animals as living things no longer work. When we consider the microscopic world, and entities that are neither animals nor plants, making an unambiguous distinction between the living and non-living becomes even more tricky.

To tackle this, biologists seek to generate a list of the properties of life that will cover everything from a bacterium to a baobab to a buffalo. The properties usually included on such a list are as follows: cellular organization; the ability to reproduce or self-replicate; growing and developing over time; using energy; being able to regulate its chemical processes (homeostasis); responding to changes in its environment; and being able to adapt to change over time, through successive generations. We can see all of these in action in animals and plants, and fungi too. Scale down to single-celled eukaryotes, and the list holds true as well—they only have one cell, but that cell shows "cellular organization" because it contains an array of smaller structures (organelles) that between them carry out all of the functions that are needed to keep it working, in a similar way to how organs and tissues work in a multicellular organism. Our single-celled eukaryote also shows a form of growth and development, as it builds its various proteins only at certain times, as needed, and has processes to repair damage.

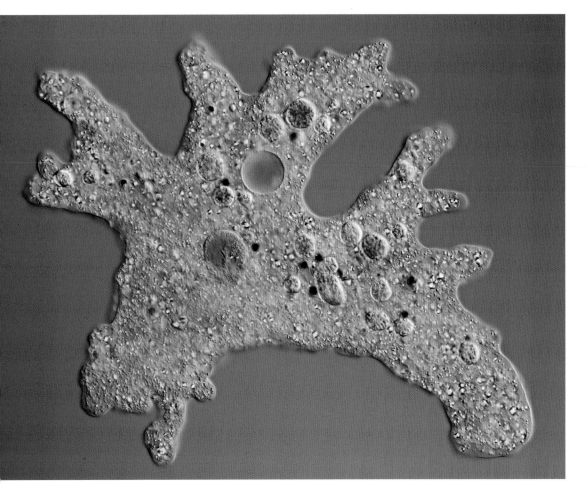

Left *We might think of an amoeba as a very simple life-form, but a powerful microscope reveals quite a complex structure, with multiple organelles and elaborate pseudopodia.*

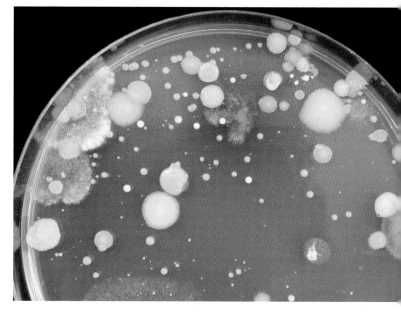

Left *Every tree owes its photosynthesizing power to the existence of simple prokaryotic cyanobacteria, now co-opted as chloroplasts within its cells.*

Below *The colonies of micro-organisms growing on a petri dish comprise millions of individual cells.*

The question of whether a prokaryote, such as a bacterium, is alive by these criteria is a little trickier, again because of the "cellular organization" factor. A bacterium is a cell but doesn't contain the numerous organelles of a eukaryotic cell—it is much simpler. However, it does still have distinct structures with separate functions, including a cell wall and membrane to hold it together, ribosomes to build its proteins, and its DNA. A virus, though, only fulfills some of the essential properties of life. While it can reproduce (within its host), and its population can certainly adapt to change through natural selection, it is not a cell (as it has no cell membrane and no organelles), which means it does not grow, develop, or exhibit homeostasis. It also does not use energy or respond to stimuli. Consequently, viruses are usually considered to be non-living, but they certainly *are* biological, straddling some kind of imaginary boundary between chemistry and life.

// Structure of viruses

As we have seen, viruses are not generally considered to be living things. Their molecular make-up, though, is unmistakeably biological, involving the same kinds of molecules (proteins, fats, and nucleic acids) from which our own cells are built. Their simple biology co-opts the much more elaborate biological machinery of true cells, both prokaryotic and eukaryotic, to carry out their self-replication. By many measures, viruses are the most efficient and successful of all biological entities on our planet and, for us, understanding how they are built and how they function is much more than a matter of scientific curiosity—viruses, unthinking, unfeeling, and unliving though they are, count among our most formidable enemies.

Below *A cow being tested for the prion disease bovine spongiform encephalopathy.*

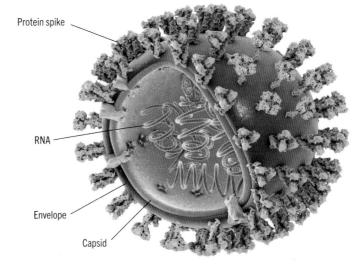

Protein spike

RNA

Envelope

Capsid

Above *Structure of a virion.*

Human red blood cell
8000 nm

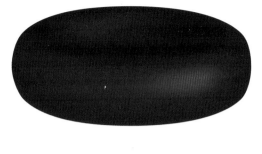

E. coli bacterium
2000 nm

Influenza virion
80–120 nm

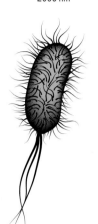

Above *Comparative size of a virion compared to a human blood cell and a bacterium.*

A single complete virus particle is known as a virion, and on average it measures about 100 nanometers in diameter (that's 0.0001 mm—by comparison, the average sphere-shaped bacterium is about 0.001 mm across and the average human cell measures between 0.01 and 0.1mm). This means that viruses are too small for us to see with a light microscope, and we have to use electron microscopy to see them. The simplest kind of virion comprises a single strand of a nucleic acid (RNA in most cases but DNA in some virus types), wrapped in a protein coat or capsid. There may also be an envelope of fat and protein molecules, derived from the membrane of the virus's host cell, around the capsid.

The shape of the virion, and the nature of its capsid and (if present) envelope, shows a lot of variation, depending on how exactly the virus is adapted to attach to its host cell, and get its nucleic acid inside that cell in order to reproduce itself. For example, a virion of the virus most of us know best at the present time, COVID-19, has a relatively large, spherical shape with club-shaped spike proteins on its envelope, which project outwards as a sort of halo or crown (giving rise to the name "coronavirus" for this type of structure) and are key to the virion being able to bind to its host cell. Host cells naturally evolve defenses to viral attack, which in turn drives the natural selection process for viruses to evolve ways to overcome those defenses, leading to the great structural and functional variety we see among the millions of distinct viral types.

Below *Structure of the prion protein which, in humans, is associated with Creutzfeldt-Jakob disease.*

SATELLITES AND OTHER SUBVIRAL AGENTS

Viroids, as we have seen, are simpler even than viruses, so are classed as subviral agents—agents because they have a kind of independent existence or agency, even though they are not living. Other subviral agents include satellites, which are virus-like entities that can self-replicate in host cells, but need a "helper virus" to do this, as well as the host cell itself.

Prions are another category of sub-viral agents. A prion is an abnormal protein that has developed infectious properties—its presence triggers other, normal proteins to form in an abnormal way. Creutzfeldt-Jakob disease is a fatal brain condition caused by a prion, as is fatal familial insomnia, victims of which become progressively less able to sleep. Prion diseases are incurable and deadly, but thankfully very rare.

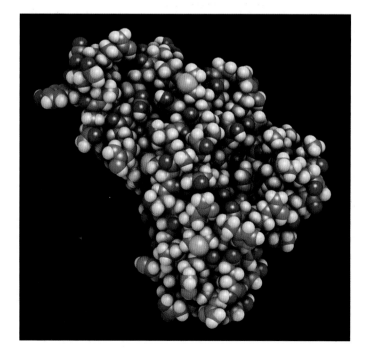

Types of viruses

We classify living things according to a nested hierarchy. All life is divided into three fundamental domains, which are Bacteria, Archaea, and Eukarya. Within the domain Eukarya there are a number of kingdoms, including Animalia (all animals) and Plantae. Each of these has subgroups called phyla nested within them—our phylum is Chordata (animals with a spinal cord). The system continues, with several classes nested within phyla, orders within classes, families within orders, and genera within families. The basic unit of classification is the species—defined along the lines of "a population of organisms that look and behave in more or less the same way and can reproduce freely with each other." One or more species make up a genus (for example, the species *Felis catus*, or domestic cat, shares its genus *Felis* with *Felis margarita*, the sand cat; *Felis chaus*, the jungle cat; and several others).

Viruses, as non-living biological things, fall outside this system but we still classify them in the same way, although the top level is called realm rather than domain. The realms are defined by fundamental traits, such as whether the virus's genetic material is DNA or RNA, and the types of proteins its genome is coded to build. Realms are then subdivided into kingdoms, and so on in the same way described above. A species of virus is defined as a population that has many traits in common, forms a replicating lineage, and affects the same host in a similar way.

Below *This chaffinch's feet are affected by* Fringilla papillomavirus. *The papillomaviruses cause warts, verrucas, and similar growths in many different animal species.*

About six realms of viruses are currently recognized by science. Among them are *Duplodnaviria*, which have a DNA genome, and include elaborately shaped, "tailed" viruses that attack bacteria, and *Riboviria*, a very large group of viruses with an RNA genome, which are round, icosahedral (forming a 3D geometric shape with 20 sides) or filament-shaped, and mostly attack eukaryotes. This latter realm includes the coronaviruses, and many other species that cause important viral diseases in humans, including flu, Ebola, and polio.

GIANT VIRUSES

Most viruses, as we have seen, are about 100 nanometers across. The giant viruses, though, belonging to the realm *Varidnaviria*, are up to four times that size, and their size is further increased by an outer layer of protein fibres up to 100 nanometers thick. They have a DNA genome that is much larger than those of other viruses (holding more than 1,000 genes in some cases, whereas most viruses hold somewhere between 10 and 200 genes on their genomes). The first giant virus to be described was initially mistaken for a type of bacteria, due to its size. Giant viruses form a "virus factory" within the cytoplasm of their host cell, rather than entering the nucleus as other viruses do. It has also been found that these "virus factories" can themselves be infected by smaller viruses (known as virophages).

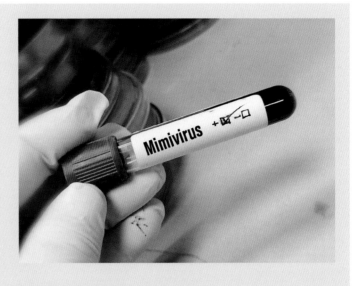

Above Mimivirus *is a genus of giant viruses.*

Below *Virus of the genus* Ebolavirus, *responsible for the severe disease Ebola.*

How viruses work

You shared a long car ride with your friend, and the conversation was flowing. You had a great time, but the next day your friend is complaining of a sore throat and stuffy nose. You experience a certain sinking feeling. You remember to take some vitamin C, and maybe you try an over-the-counter cold prevention remedy, but you don't hold out much hope. Sure enough, two days later you wake with those same symptoms. On that car ride, you and your friend shared more than just words, and now you have several billion uninvited guests in your upper respiratory tract.

As we have seen, viruses are not cells as such, and they exhibit some of the properties of true living things, but not all. Their means of reproduction is a halfway house in itself—they do reproduce themselves (prodigiously) but need to co-opt living cells (either prokaryote or more complex eukaryote cells) to accomplish this. Cells contain all the equipment needed to assemble chains of RNA and DNA, and an invading virion hijacks this machinery so that the cell produces new viral RNA or DNA instead.

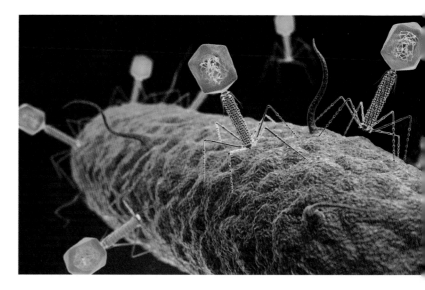

Above *Illustration of bacteriaphage viruses attacking a bacterial cell.*

Below *How a virus invades a cell and co-opts its RNA replication system.*

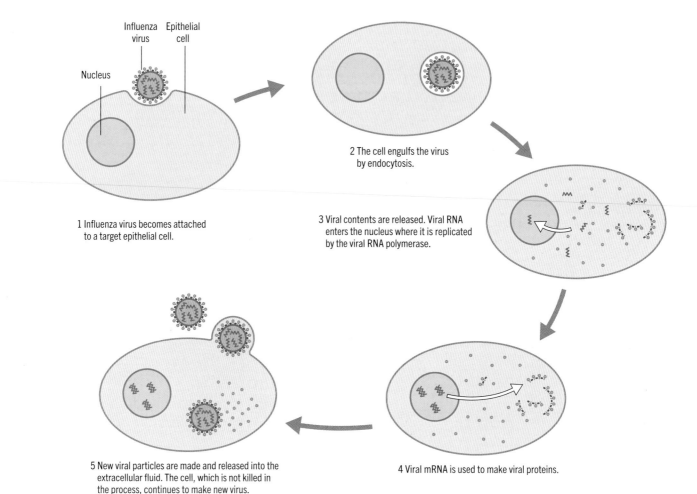

Nucleus
Influenza virus
Epithelial cell

1 Influenza virus becomes attached to a target epithelial cell.

2 The cell engulfs the virus by endocytosis.

3 Viral contents are released. Viral RNA enters the nucleus where it is replicated by the viral RNA polymerase.

4 Viral mRNA is used to make viral proteins.

5 New viral particles are made and released into the extracellular fluid. The cell, which is not killed in the process, continues to make new virus.

The process by which a virus replicates within a host cell has six key stages:

1 **Attachment** A virion attaches to a host cell, typically through proteins on its envelope binding to a specific point on the outer surface of the cell.
2 **Penetration** The envelope of the virion and the membrane of the cell fuse together so the contents of the virion can enter the cell. (In some cases, the entire, intact virion is engulfed by the cell.)
3 **Uncoating** The virion's inner capsid is broken down, freeing its RNA (or DNA). (Some bacteriophages remain attached to the host cell but inject their genetic material inside.)
4 **Replication** The viral RNA or DNA is replicated in the cell (usually within the nucleus) and the host cell also builds new viral proteins, using the viral messenger RNA. The exact mechanism varies by the type of virus and host, but is triggered and guided by the virion's enzymes, with the cell's own processes halting.
5 **Assembly** Entire new virions are built from the newly generated RNA or DNA, and the viral proteins.
6 **Release** The new virions are released from the cell, either by the cell dying and breaking down, or by the virions migrating to the cell membrane, and using a part of the membrane to form their own envelope.

Above *Many viruses that cause infection of the respiratory tract are spread in microdroplets released when a sufferer coughs or sneezes.*

Viruses need to travel within a host's body (or between hosts) and have no way to move on their own. However, the actions of their host can help them to get around. The virus that you caught from your friend in the car traveled between you as a hitch-hiker, within minuscule droplets of fluid that were exhaled by your friend, and then inhaled by you. Sometimes they must rely on random encounters with the host—for example, bacteriophages moving within a liquid substrate that also contains host bacteria. A virion that attaches to the flagellum (a tail-like filament, used for movement) of a bacterium may move along the flagellum to reach the bacterium via a corkscrewing motion as the flagellum rotates, because of the spiral shape of its attachment protein.

// Viruses—abundance and diversity

We like to think of ourselves as the dominant lifeform on this planet, and, looking at things on a human scale, that does indeed seem to be the case. However, the beetles would take issue with this—with 400,000 described species, they make up 25 percent of all known animal species. So would the ants—there are 2.5 million of them for every one of us. Prokaryote abundance easily beats that of any eukaryote, with 5×10^{30} individual Bacteria and Archaea cells on the planet. However, viruses—even though they require non-viral hosts to reproduce—are ten times more abundant even than the prokaryotes. Your own body is home to an alarming 380 trillion or so individual virions at all times.

Luckily for us (and all other living things), viruses are generally highly host-specific, so only a tiny proportion of them will affect any given species, including *Homo sapiens*, and most are either harmless or cause very mild symptoms. Some others are even thought to have beneficial effects. Out of those 380 trillion virions on and inside you (your personal "virome"), a sizeable proportion are not actually infecting your own cells but instead invade the cells of the various bacteria that are also living on and in you—and they could be harming or helping those bacterial hosts, too. Nor is your viral population the same as that of other humans—in fact, even the people you live with will share only about 25 percent of the virus species that make up your virome.

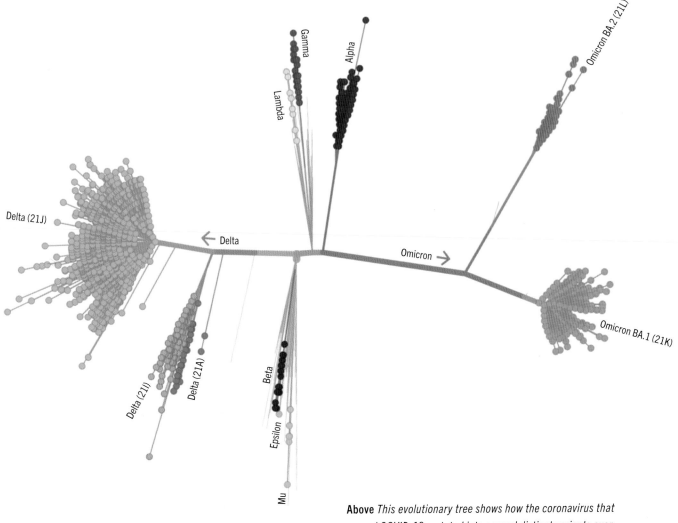

Above *This evolutionary tree shows how the coronavirus that caused COVID-19 mutated into several distinct variants over a very short space of time.*

Viruses need hosts, but they can be found everywhere on our planet—in the soil, in water, in the air. Being non-living, they cannot die but can be deactivated, and this may happen within hours when they are not inside their host. However, some can persist and remain capable of infecting a host for a long time—scientists in 2014 were able to reactivate samples of the giant *Pithovirus* that were found in 30,000-year-old Siberian permafrost. We will never know the total number of distinct virus species in the world, but it is likely that every one of the 1.74 million or so known species of eukaryotes each has at least 20 uniquely host-specific viruses. That 1.74 million is nowhere near all of the world's eukaryotes, either—a recent estimate says that there are probably another 7 million awaiting discovery. Add the viruses that affect the prokaryotes, and we are in the realms of very big numbers indeed.

THE ORIGINS OF VIRUSES

As with viroids, it is possible that viruses have existed for longer than true living organisms and descend from the RNA molecules that inhabited RNA World. If this were the case, then once life evolved, these RNA molecules evolved into forms that could invade living cells. Two other theories are also proposed. One, the progressive hypothesis, suggests that viruses evolved from small bits of genetic material (RNA and DNA) in living cells, which evolved the ability to move between cells. The other is the regressive hypothesis, which suggests viruses are "degraded" or simplified forms of living cells, which were originally parasites of other cells and became reliant on those cells for their own replication. Some support for this latter idea comes from the discovery that certain bacteria, like viruses, cannot replicate without help from the host cells that they infect, but did descend from ancestors that were free-living.

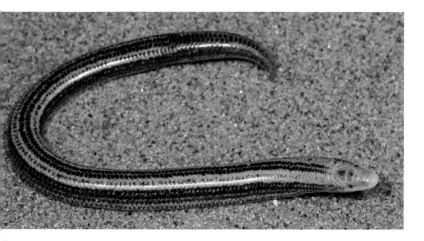

Above *Evolutionary adaptation can sometimes mean a loss of certain functions, such as the disappearance of limbs in some types of burrowing lizards.*

Right *The giant* Pithovirus, *revived from ancient Arctic permafrost.*

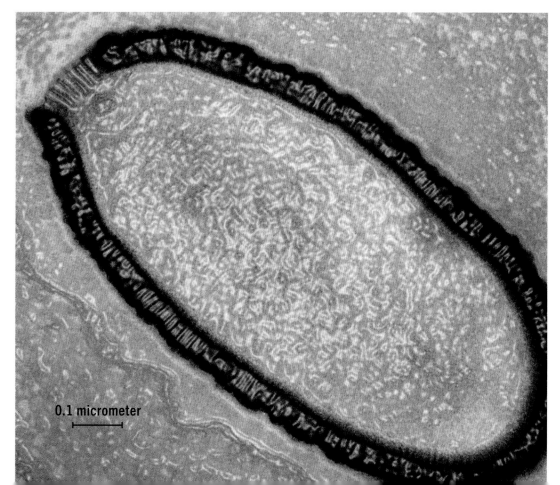

0.1 micrometer

// Antiviral activity

Animals, plants, and other living things are all subject to viral infection, and this may in some cases harm or kill them, but they are not defenseless. There are certain things that animals in particular can do in terms of behavior to help reduce their exposure to viruses, such as avoiding overcrowded situations where spreading and catching diseases happens more easily, and many animals do this naturally. There are also natural physical defenses, such as the waxy cuticle on plant leaves, and the outer layer of skin of animals, which are there to protect from injury and help ensure the cells below don't lose too much water, but also to help keep disease-causing agents (pathogens). Once a virus gets into the system and starts to attack cells though, combat takes place on the microscopic scale.

In humans and other mammals, there are two components to our immune systems, both of which can help us tackle viruses. The innate immune system includes the physical barriers mentioned above, but also white blood cells called macrophages that detect, attack, and neutralize pathogens, as well as the body's own cells if they have become infected. Macrophages and some other types of white blood cells interact with B and T cells, which form the second component of our immune system, the adaptive immune response.

Below *Several drug companies produced effective but distinctly different vaccines against COVID-19.*

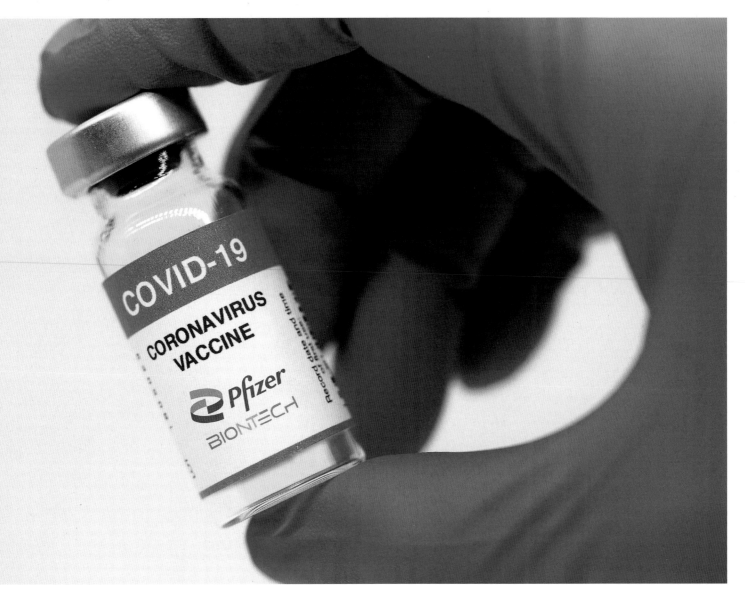

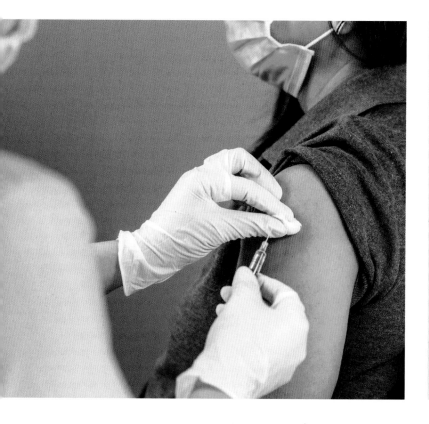

VACCINATION

A great many humans have received at least one dose of a vaccine designed to combat the COVID-19 virus since the disease emerged in late 2019. These vaccines work by safely teaching the adaptive immune system to recognize the antigen of the target virus, so that when a real infection occurs, the necessary antibodies have already been made and are circulating in your blood, ready to go. In the case of the various COVID-19 vaccines, some contain genetically engineered messenger RNA, which makes your cells manufacture the COVID-19 antigen. Your immune cells then create antibodies to that antigen. Some other COVID-19 vaccines use actual COVID-19 genes placed inside a different, non-dangerous virus. When this virus enters cells, the cells respond by making the COVID-19 spike protein, which triggers your B cells to make antibodies. Other forms of vaccines, devised to combat viral and also bacterial infections, may involve giving the patient a very small dose of the live pathogen, or a dose of deactivated pathogen.

Above *The speed at which the COVID-19 vaccine was devised and delivered from 2020 was unprecedented.*

B cells produce antibodies, large protein molecules that bind to one particular molecule (an antigen) on the surface of a virus or other pathogen. After your immune system has encountered a particular virus for the first time, your B cells "remember" how to make the right antibody to fit that virus for a long time (for a lifetime in some cases). Antibodies begin life attached to B cells, but then circulate in the blood on their own. Once antibodies have bound to the outside of a virus, cells containing that virus are "marked for destruction." T cells come in different types, some of which work with B cells to help to accelerate the antibody-creation process, while others will kill cells that have become infected.

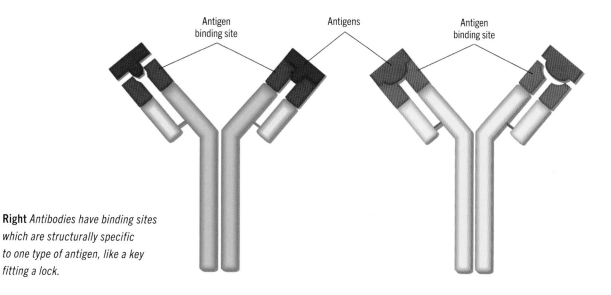

Right *Antibodies have binding sites which are structurally specific to one type of antigen, like a key fitting a lock.*

PROKARYOTES

Most biologists consider there to be three fundamental domains of life. We, along with all other animals, plus plants, fungi and other complex life, make up just one of those domains. The other two, Bacteria and Archaea, are prokaryotes—simple but fully living organisms which have existed on Earth for billions of years. Their diversity of form and lifestyle allows them to exploit every habitat in the biosphere, and some can thrive in extreme conditions that no plant or animal could endure. Through co-operative living with each other, they gave rise to complex life, and today many of them make their living in co-operative or parasitic relationships with complex organisms, including ourselves.

The simplest forms of life on our world, the prokaryotes are also by far the most successful and numerous.

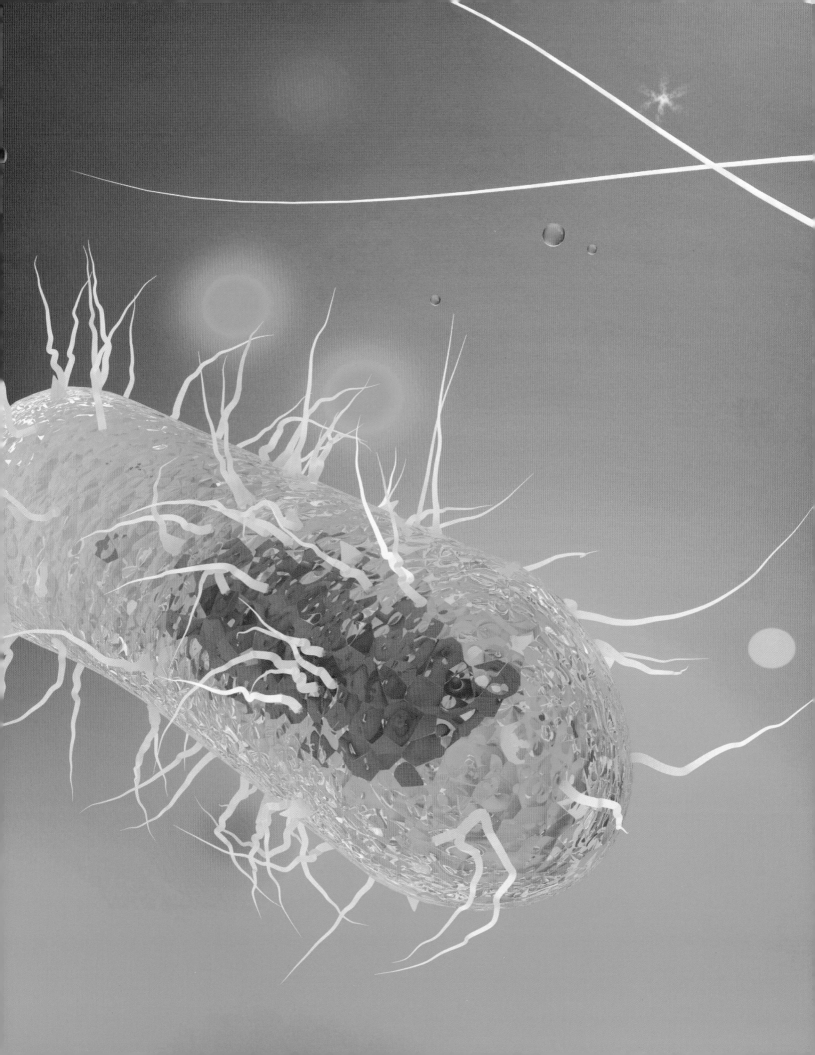

// What defines a prokaryote?

The vast majority of life on Earth is comprised of prokaryotic organisms. When life was classified into several fundamental kingdoms, all prokaryotes were grouped into one of these—the kingdom Monera. Now that our knowledge has expanded, we define the most fundamental division of life as the domain (with domains further subdivided into kingdoms). Two of life's three domains are prokaryotic—the Bacteria and the Archaea, with complex life making up the third (Eukaryota). To us, Bacteria and Archaea seem extremely similar—both are tiny single-celled organisms and their cells are much smaller and simpler than the cells that form our own bodies. However, in terms of their biology and biochemistry, they are as distinct from one another as either of them is from eukaryotic life. In fact, it is Eukaryota and Archaea that are the most similar and the closest cousins in terms of evolutionary relationships, with Bacteria the outlier.

Prokaryotes are, on average, a tenth the size of eukaryotic cells, and have a less complex structure, but some eukaryotic cells are much smaller and simpler than others. The fundamental defining point that separates a prokaryote from a eukaryote is that it has no cell nucleus. In a eukaryotic cell, the chromosomes holding the DNA are contained within a membrane-bound cell nucleus. In a prokaryote, the solitary DNA chromosome is held within the cell's cytoplasm. Some highly specialized eukaryotic cells also lack a nucleus—our own red blood cells are an example—but only mature red blood cells have lost their nuclei (and with it, all of their DNA, along with their ability to make proteins and RNA). The precursor cells from which they develop do have nuclei.

As well as lacking a nucleus, prokaryotes also lack most of the other membrane-bound cell organelles that you will find in most eukaryotic cells, such as centrioles, endoplasmic reticulum, the Golgi complex and mitochondria, and, in plants, chloroplasts. In fact, the mitochondria and chloroplasts that carry out (respectively) energy generation and photosynthesis for eukaryotic cells are highly likely to have originated as prokaryotic Bacteria, which came to live co-operatively within other, possibly Archaean prokaryotes. See page 70 for more on this theory of endosymbiosis.

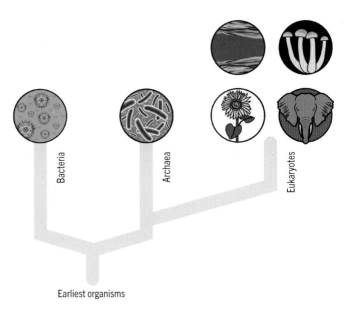

Above *The three-domain classification system of life. Eukaryotes are generally considered to have emerged from within Archaea, although Bacteria also played a crucial role in their evolution by contributing mitochrondria and plastids.*

CLASSIFICATION CONUNDRUMS

The Archaea, when first discovered in the 1970s to be somewhat distinct as a group, were treated as a subgroup of the prokaryote kingdom Monera, and were given the name Archaebacteria, while the group containing all other bacteria was called Eubacteria. Today, their traits and distinctiveness, and their closer relationship to Eukaryota, are well understood, and indeed some biologists consider there to be just two, rather than three, domains of life—Bacteria and Archaea, with Eukaryota relegated to a subgroup that evolved within an already established Archaea, rather than from a common ancestor shared with both Bacteria and Archaea. Back in the early 20th century, prior to advanced microscopy technology, we naturally did not accord the prokaryotes this (correct) level of importance. A popular way to classify life then involved four or five kingdoms as the top-level division, those kingdoms being Protista (single-celled eukaryotes), Plantae (plants, with fungi included or considered a separate kingdom in a five-kingdom system), Animalia (animals), and Monera (all prokaryotes). As we have looked more closely at smaller and smaller entities, the significance of our own eukaryote lineage has also diminished in size as our understanding of prokaryotic life has grown.

Below *The fundamental divisions of life according to the old "five-kingdom" system, now generally replaced by the three-domain system.*

ANIMAL

Cnidaria

Porfera

Platyhelminthes

Molluscs

Annelids

Echinoderms

Arachnids

Crustaceans

Amphibians

Fish

Insects

Reptiles

Birds

Mammals

PLANT

Equiseta

Angiosperms

Gymnosperms

Mosses

Ferns

FUNGI

Ascomycetes

Basidomycetes

PROTISTA

Red algae

Ciliated protozoa

Green algae

Brown algae

Flagellated protozoa

Amoeboid protozoa

MONERA

Archaebacteria

Eubacteria

// Structure of a typical prokaryote

As we have seen, the prokaryotes are structurally simple compared to eukaryotes, though much more complex than the not-quite-living viruses and other entities that we explored in Chapter 1. Nevertheless, within their tiny cells they possess all the machinery to fulfill the criteria of life, and indeed many are highly successful and prolific organisms that are found almost everywhere on Earth. Others are rare, localized, and highly specialized but adapted to survive and thrive in some of the most extreme conditions on our planet, coping with temperatures, radiation levels, or concentrations of potentially harmful chemical compounds that no eukaryote could endure.

Although they lack nuclei and most of the other organelles found in eukaryotes, prokaryotes do have ribosomes, the structures responsible for building proteins according to the genetic code held on the chromosome. They also have an outer cell membrane, which contains their contents while allowing control of the movements of molecules into and out of the cell. Most types have a more rigid cell wall outside of the cell membrane, which provides protection and maintains

Below *Structure of a prokaryotic cell.*

Above *The curious shape and arrangement of* Haloquadratum walsbyi *cells.*

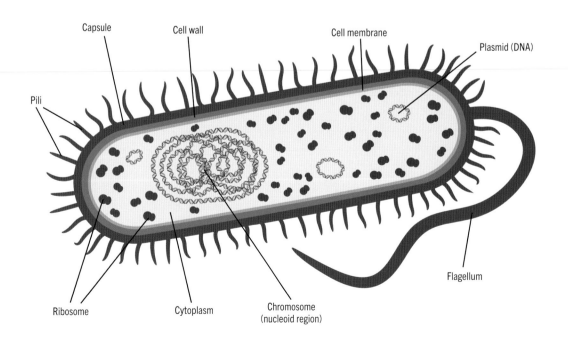

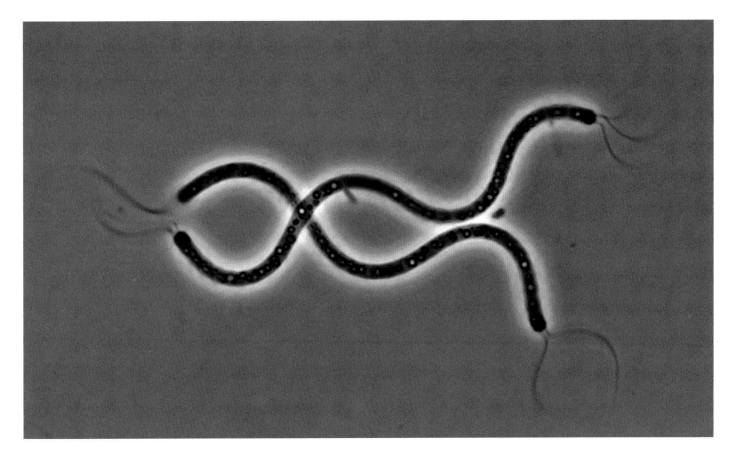

the cell shape. Some rod-shaped prokaryotes also have a flagellum at one or both ends—a whip-like "tail" that allows them to move more easily and swiftly, and some have pili, which are hair-like strands of protein that help the cells stick to each other or other surfaces when needed.

The interior of the cell contains cytoplasm, the same watery gel-like substance that fills eukaryotic cells and carries a range of chemical compounds, including enzymes, nutrients, and waste materials. The chromosome is held within a nucleoid, which is a distinct but non-membranous region of the cytoplasm. In some bacteria species, the cytoplasm may also contain one or more plasmids, which are very small, usually circular sections of DNA. The genetic code in a plasmid usually allows the cell to produce an additional protein, which may have some helpful functionality, for example allowing the cell to break down a toxin, but in many cases the plasmid DNA has no apparent function. Plasmid DNA can also sometimes combine with the cell's main chromosome.

Above *Flagellae, used for propulsion, are present in several lineages of prokaryotes, both Archaea and Bacteria.*

PROKARYOTE SHAPES

The most familiar prokaryote shape is the rod, or bacillus (plural bacilli). These organisms are shaped like straight sausages, with a cylindrical shape and rounded ends. *Bacillus* is also the name of a particular bacterial genus. Most other prokaryotes are spherical or egg-shaped—this shape is known as coccus (plural cocci). A few others are shaped like a spiralled coil, a form known as spirillum (plural spirilla). Members of the Bacteria genus *Vibrio* have a curved rod shape and a flagellum, giving them a comma-like shape, while the aquatic Archaea species *Haloquadratum walsbyi* has flat, square cells that join together to form thin, floating sheets.

// Classification of prokaryotes

The prevailing modern view of prokaryotes as forming two of the three domains of life was arrived at over several decades, via a circuitous route. Thanks to knowledge acquired using improved microscopy technology, we have arrived at a classification that is broadly agreed within the scientific community. It would be a mistake, though, to assume that this journey of discovery is over now— there remains much to learn about both prokaryotes and eukaryotes, and new information could easily overturn today's system in years to come. However, it has become increasingly clear that the Archaea and Bacteria, despite their superficial similarities, are in fact very different, and any future revision of their classification seems very likely to preserve their separation on the basis of this distinctiveness.

The earliest attempts to classify prokaryotes treated them as a single lineage, and biologists subdivided them into groups based on traits such as their cell shape, and the types of material they consumed. By the 1970s, there were still very few examples of Archaea known to science, but one group that was well studied were the methane-producing organisms that occur in places such as marshy swamps, cow stomachs, and human colons. Wherever they live, they produce odoriferous methane as a metabolic by-product, which is then bubbled, burped, or farted out. Decoding the ribosomal RNA genes of these organisms revealed them to be substantially different to those found in bacteria, and the two groups were reclassified on this basis. The differences between Archaea and Bacteria that we now know of include their wider genetic make-up, the enzymes they use when making RNA and proteins, and also the way that they utilize compounds in their structure—for example, the fatty acid-based molecules in their cell membranes are of distinctly different types.

Biologists divide both Archaea and Bacteria into several taxonomic groups or clades, which themselves are further divided into smaller groups based on shared characteristics. However, attempts to classify them by the rather rigid kingdom-phylum-class-order-family-genus-species system that we use for eukaryotes does not work very well. One of the reasons for this is that, as they evolve, different lineages may swap genes with one another (horizontal gene transfer), confusing the tidy taxonomic tree that we are trying to build. We also need to consider that, in all likelihood, the number of species in both groups that we've discovered and described so far is a small proportion of the total. All of that notwithstanding, many groups and subgroups of both Archaea and Bacteria have been classified on the basis of their genes, which reveals how they have evolved and diverged over the millennia.

Below *A bubbly, methane-generating swamp, habitat for certain Archaea.*

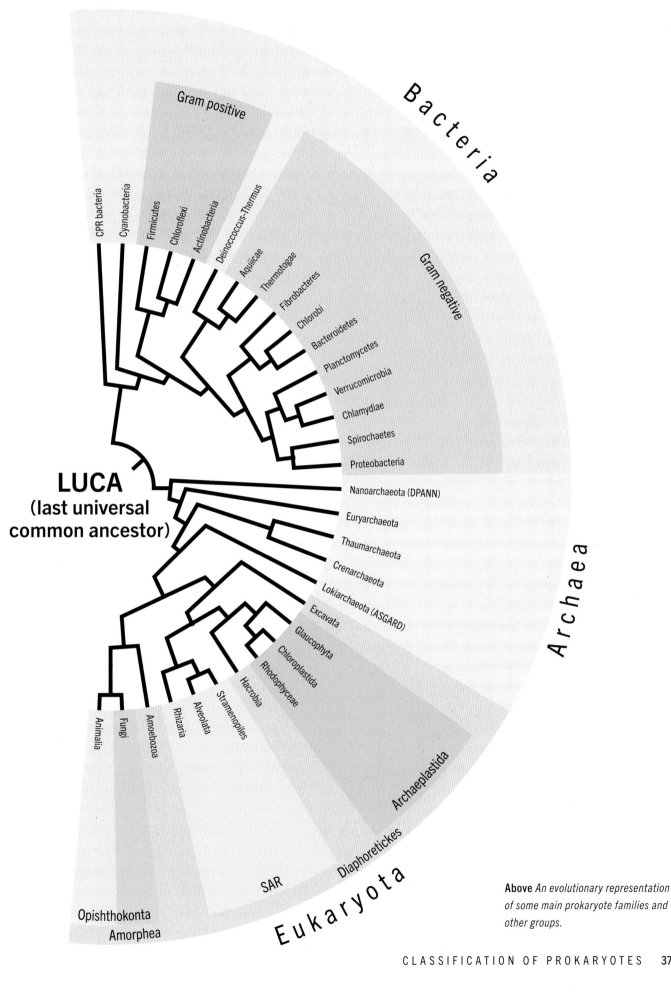

Gram positive

Bacteria

CPR bacteria
Cyanobacteria
Firmicutes
Chloroflexi
Actinobacteria
Deinococcus-Thermus
Aquiicae
Thermotogae
Fibrobacteres
Chlorobi
Bacteroidetes
Planctomycetes
Verrucomicrobia
Chlamydiae
Spirochaetes
Proteobacteria

Gram negative

LUCA
(last universal
common ancestor)

Nanoarchaeota (DPANN)
Euryarchaeota
Thaumarchaeota
Crenarchaeota
Lokiarchaeota (ASGARD)

Archaea

Excavata
Glaucophyta
Chloroplastida
Rhodophyceae
Hacrobia
Stramenopiles
Alveolata
Rhizaria
Amoebozoa
Fungi
Animalia

Archaeplastida

Diaphoretickes

SAR

Opishthokonta
Amorphea

Eukaryota

Above *An evolutionary representation
of some main prokaryote families and
other groups.*

// Archaea—characteristics

The rainbow shores and deep blue waters of the Grand Prismatic Spring in Yellowstone are stunningly colorful. However, that water is close to boiling temperature and so for safety is best admired from a distance for us eukaryotes. Yet this environment, highly inhospitable to us, is home to a variety of prokaryotes, and is the place where some of the first known Archaea were discovered. Other early finds of Archaea also took place in various hostile-seeming environments, and so the Archaea are often thought of as extremophiles, adapted to live where few other organisms can. Today, we know better, as we have found them in many other environment types.

The first distinctive Archaea traits were described by Carl Woese and George E. Fox in 1977. These microbiologists' work began with comparing the genes in the ribosomal RNA of Archaea with those of Bacteria, discovering a significant difference. Later, they identified other key differences—that Archaea lack the structural molecule peptidoglycan in their cell walls, and that they possess two enzymes not found in Bacteria. These discoveries led in due course to the proposal that life should be reclassified into the three domains of Archaea, Bacteria and Eukaryota or Eukarya—a revision so dramatic and (in some quarters) so controversial that it was not accepted by mainstream biology for a decade, despite robust supporting evidence. This leap forward in our understanding of microbiology and the fundamental categories of living things is still known as the "Woesian Revolution."

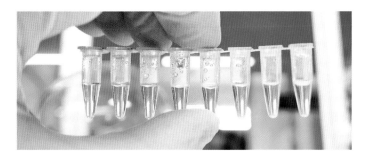

Above *Lab tests such as the polymerase chain reaction have enabled us to learn much more about Archaea populations and diversity.*

Opposite *The Grand Prismatic Spring in Yellowstone National Park, Wyoming, owes its amazing colors to microbial mats of Archaean and Bacterial life.*

Archaea overlap with Bacteria in many of the traits that we can observe most easily—they come in the same range of shapes and sizes and their cells are made up from similar components. However, genetic and molecular differences are more important and accurate when it comes to correct classification. The same issues can be seen in other groups of living things—we see the similarities between fish and dolphins, for example, and may once have assumed they must be close relatives on this basis, but a deeper and closer look shows how different they are. Unrelated organisms with similar ways of life often evolve similar traits (convergent evolution), and similarities between some Archaea and Bacteria groups are likely to be examples of this.

Above *Microbial mats are known to form in all hot springs around the world. The species they hold are enduring near-boiling temperatures and strongly alkaline or acidic conditions.*

FIRST FIND YOUR ARCHAEA

The relatively recent flurry of discoveries of Archaea in "non-extremophile" habitats is owed to a technique called the polymerase chain reaction. Invented by the American biochemist Kary Mullis in 1983, it is a way of quickly multiplying a minuscule DNA sample many times over, creating numerous copies of the sample, and thereby providing enough material for practical study. The necessary replicating reaction is set up by adding a DNA polymerase enzyme to the sample, and then set in motion by applying heat and then cooling, to stimulate the enzyme to do its work and help the DNA to replicate. Through PCR, the prokaryotes present in samples of soil, seawater sediment and animal and plant tissues can be discovered and categorized.

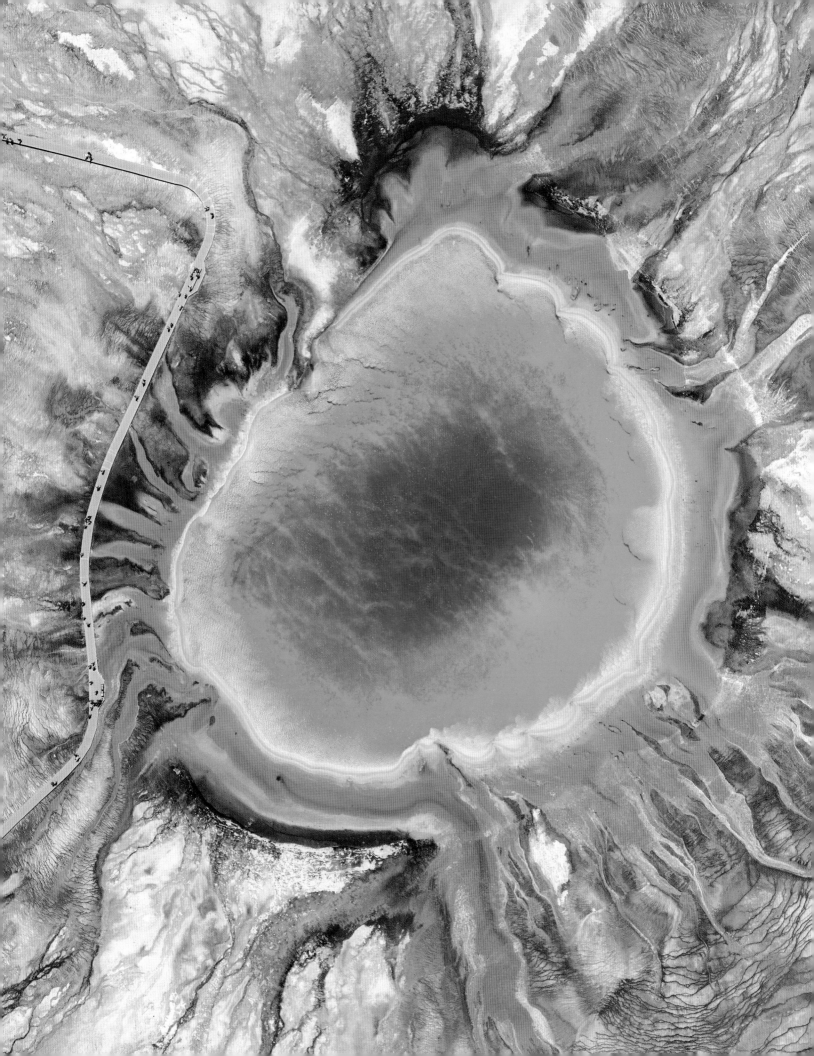

// Archaea—types

It's likely that the classification of Archaea will go through lots of changes over the coming years, as we are still at an early stage when it comes to exploring and categorizing this group of organisms, and there is no general agreement within the scientific community at this time. That said, most microbiologists currently recognize two or more distinct fundamental groups or phyla within Archaea, each of which has its own distinctive characteristics.

Thermoproteota, also known as Crenarchaeota, are a large and diverse group of Archaea, though they share the same type of ribosomal RNA, and also, unlike other Archaea, they usually lack histone proteins, which are a type of protein found in chromosomes. Most are found in underwater habitats and are particularly abundant (probably more so than any other microorganisms) in the seabed, but they have also been detected in freshwater habitats and soil. Most are adapted to survive in high temperatures—among

them the genus *Sulfolobus*, which can, when necessary, use sulphur rather than oxygen in its energy-producing metabolic pathway (anaerobic respiration). The very ancient genus *Thermoproteus* always uses anaerobic respiration, its lineage dating back to a time in Earth's history when oxygen was not as readily available as it is now.

Euarchaeota are another diverse group, which come in rod and sphere shapes, and differ from other groups in the make-up of their ribosomal RNA and their DNA-building enzyme. They include a variety of methane-generating species, which live in various temperature conditions, ranging from volcanically heated springs to Arctic ice, as well as in human and other animal digestive tracts. This group also holds the salt-loving or halophilic Archaea, which include some very ancient lineages. Archaean DNA extracted from fossil material more than 100 million years old is genetically extremely similar to the living species *Halobacterium salinarum*.

Below *This Australian salt lake owes its pink color to the presence of the Archaean* Halobacterium.

Above *This map shows the Gakkel Ridge, in the Arctic Ocean, location of many hydrothermal vents that provide habitat for extremophile Archaea.*

Korarchaeota is a group of Archaea known from hydrothermal environments, varying in the exact temperatures and levels of salinity they prefer. Their cells are long, narrow rods. Nitrososphaerota, another heat-loving group, are similar in ecology to members of Thermoproteota and were formerly classified within that group, but we now know that their cells contain certain enzymes and other compounds that are not found in any other Archaea.

Nanoarchaeota has only one known representative at present, but this species, *Nanoarchaeum equitans*, found in an Arctic hydrothermal vent in 2002, is distinct from other groups. For one thing, it is exceptionally tiny in size and possesses the shortest genome of any known organism. Its ribosomal RNA, the structure of which is used to differentiate fundamental Archaean groups, is also unique.

ORIGIN OF EUKARYOTES

Another proposed fundamental group of Archaea, the Asgardarchaeota or just Asgard, is especially interesting because it is thought by some microbiologists to be the lineage within which Eukaryota arose (as well as several other Archaea groups). Members of Asgard share a number of protein types and genetic traits with eukaryotes. One Asgard species has recently been found to have close symbiotic links with two Bacteria species. This may reflect the theorized endosymbiotic origin of the first eukaryotes—Bacteria and Archaea species living in close and interdependent proximity, with the Bacteria eventually living as energy-generating mitochondria within enlarged Archaea cells (see page 70 for more details on endosymbiosis).

// Archaea—abundance and distribution

As we have seen, Archaea are heavily outnumbered by Bacteria on Earth in terms of both numbers of species and numbers of individuals—as far as we know. We are, of course, yet to thoroughly explore every square centimeter of our world in search of these tiny organisms and there is much more to learn about the abundance and diversity of prokaryotes. In any case, we know Archaea to be extremely widespread and to occur everywhere in our world, from the deepest seabeds to the frozen poles.

The habitats that hold the biggest Archaean populations are marine, and they make up about 37 percent of microorganisms found on and in sediment on the seabed. In fact, the presence of anaerobic Archaea has been detected in sediment as deep as 1.5 miles (2.5 km) below the ocean floor. They are also abundant in freshwater habitats but exist on dry land too, contributing up to 10 percent of the microbial cells present in soil.

As with Bacteria, populations of some Archaea live on and inside our own bodies (though Bacteria outnumber Archaea considerably). However, one of the ecological traits that Archaea does not (so far as we know) share with Bacteria is that no Archaea species are known to be pathogenic to any other living thing. We must remember that the majority of Bacteria are also not pathogenic, and only a very small proportion of all known species are harmful to our own species. But the few pathogenic types that we know of are collectively responsible for a great deal of human suffering and death, and with antibiotic resistance a very real and growing problem for us, it's no surprise that we consider this trait to be of extreme importance.

Below *Marine life does not stop at the seabed. Archaea can live at depths within marine sediment that far exceed the depth of the water above.*

BIOMASS

When it comes to measuring the significance of different groups of organisms as part of life on Earth, biomass is a useful concept, often more so than number of individuals, or diversity of species. A big adult African elephant is just a single animal, but it weighs 4 US tonnes (4,000 kg), or 4 million grams, whereas one house mouse weighs just 1.4 oz (40 g), so we can assume that it would take an awful lot of house mice to have the same influence in their environment as the lone elephant. Biomass is often expressed as the weight of carbon that a given number of organisms collectively hold in their tissues (about 50 percent of their total mass). The Earth's total population of Archaea is estimated to contribute about 7 gigatons of carbon to our planet's biomass. This compares to a hefty 70 gigatons for Bacteria and a massive 450 gigatons for plants, but greatly outstrips the biomass of the Earth's entire animal population (just 2 gigatons).

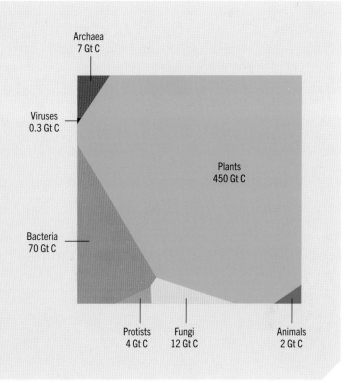

Right *The proportion of Earth's biomass that is contributed by different groups of living things.*

Below *Marine sediment varies in thickness—the deeper layers are only habitable by anaerobic organisms.*

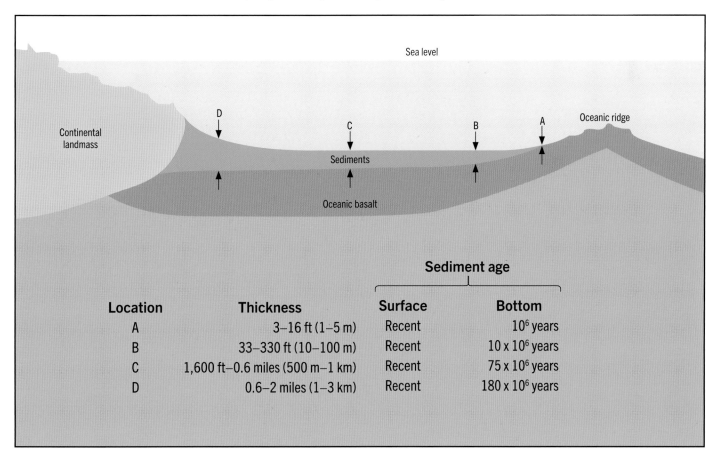

Location	Thickness	Sediment age	
		Surface	Bottom
A	3–16 ft (1–5 m)	Recent	10^6 years
B	33–330 ft (10–100 m)	Recent	10×10^6 years
C	1,600 ft–0.6 miles (500 m–1 km)	Recent	75×10^6 years
D	0.6–2 miles (1–3 km)	Recent	180×10^6 years

// Archaea—extremophiles

The Archaea as a group provide evidence of the amazing adaptability and toughness of life on our planet. They are among the most notable of the extremophiles, able to survive in the harshest and most life-unfriendly conditions, and to make use of the most unlikely resources. Another extremophile trait is to be able to enter a dormant stage, to see them through times when conditions are even more difficult, and they may survive in this almost non-living state for many years, returning to normal when conditions improve.

One of the reasons that Archaea excel as extremophiles is their very long history, which has seen them through some turbulent times. There is evidence in the form of stromatolites (ancient sedimentary formations of fossilized microorganisms) that the first Archaea appeared some 3.5 billion years ago and were perhaps the first truly living things on Earth. In the intervening years, the Earth has gone through dramatic shifts in climate and weather patterns, radiation levels from the Sun have been enormous at times, the continents have migrated across the oceans, fused, and separated again multiple times, some alarmingly large meteors have struck land and sea, and huge amounts of seismic and volcanic activity have occurred. Archaea have needed to cope with all of this, and while countless lineages have no doubt died out over the millennia, others have adapted, evolved, and survived.

Examples of the extreme conditions some Archaea exist under include, as we have seen, volcanically heated springs of near-boiling water, and also waters with an extremely high or low pH, extremely salty water (they are among the few organisms to survive in the Dead Sea, making it not quite as dead as it appears). Some remain fully active in the -4°F (-20°C) cores of Arctic sea ice in winter, and some are known to live and grow at temperatures exceeding 230°F (110°C). Others live in the deep sea and can tolerate pressures well in excess of that found in the deepest ocean trenches. The lowest air pressures on Earth, found at the peaks of big mountains, offer little challenge to some Archaea, and others can handle extreme solar radiation.

Below *The Dead Sea and its surroundings seem very bleak but an array of well-adapted microbiotic life is present here.*

OTHER EXTREMOPHILES

Extremophile means "lover of extremes," which may be a bit of a misnomer as some extremophiles are actually much happier in more gentle conditions, but what unites them is their toughness. Most extremophiles are prokaryotes, whose rapid reproduction rate gives them the edge in evolutionary adaptation to different conditions. However, a few animals qualify, too. The tardigrades, tiny, chubby eight-legged invertebrate animals also known as "water bears," are perhaps the best-known—tests show that, in a dormant state, some species can survive extremely high pressures, dehydration to the point of total desiccation, and starvation for years, and can even remain alive for some time while fully exposed to the airless vacuum and unprotected radiation of outer space. But they much prefer their normal life of wandering around in the film of water on moss or leaf litter.

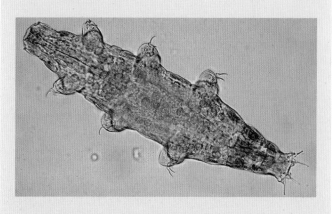

Above *Tardigrades are near-microscopic but are complex multicellular animals, with some very impressive extremophile capabilities.*

Below *Thermophilic microbial mats. The temperature of such water may regularly reach almost boiling point.*

// Bacteria—characteristics

Alongside Archaea, the domain Bacteria completes the prokaryotic life on Earth. We tend to think of them as pathogenic microbes that cause various unpleasant infectious diseases in our species and other complex living things. However, for half of the evolutionary history of life, there were no eukaryotes around for any members of Bacteria to infect. Prokaryotes were the only life that existed on our planet from 3.5 billion years ago until 1.8 billion years ago, when the first eukaryote cells appear in the fossil record. So, prokaryotes have a long history as free-living organisms, making their living in a wide range of ecological niches and not infecting anything (although they themselves were and still are attacked by viruses). Only when eukaryotes existed did a few Bacteria lineages begin to evolve a pathogenic ecology, just as some bacteria became food for tiny eukaryotes.

The amazing abundance and diversity of bacteria sets them aside as the most successful domain of life (by far). The plants may outweigh them in biomass, but in other measures bacteria are world-dominating organisms. They are highly likely to outlive all eukaryotic life as our planet continues through its life cycle (towards its inevitable destruction, when our expanding sun consumes it, some 7.6 billion years from now). Archaea may outlast Bacteria in the end, given their propensity for survival in the most unlikely and environmentally hostile settings, but plenty of bacterial species are tough extremophiles too. Some also have the ability to remain dormant for very extended periods, and a few can produce a special extra-durable endospore, which can remain dormant but viable for thousands of years.

Below *Bacteria come in a variety of shapes, although we tend to think of the classic rod- or sausage-shaped bacillus form.*

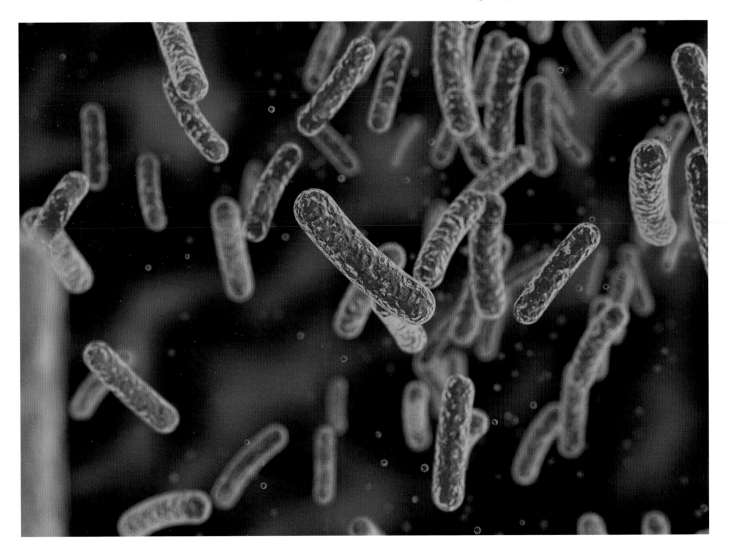

Their adaptability is very apparent to us as we do battle with some of them in our labs and hospitals. A very rapid reproductive cycle (some species are able to double their population every four minutes in the right conditions), with copying errors in their DNA replication, means that any genetic mutation that improves their survival chances will come to dominate the population very quickly. This is why antibiotic resistance in pathogenic bacteria is such a serious threat to our own species.

Bacteria have a similar cell structure to Archaea. They may be rod-shaped (bacilli), spherical (cocci) or very thin, long and spiral-shaped (spirilla), with a small number shaped like curved rods or commas (vibrio). More unusual shapes have been observed on occasion, usually in response to constraints imposed by particular living conditions. Their cytoplasm contains ribosomes and a chromosome and is bound by a cell membrane and (often) a cell wall outside of that, which may bear hair-like pili to help it stick to other cells or surfaces. Some Bacteria, especially rod-shaped and spiral forms, have flagelli to help them move around.

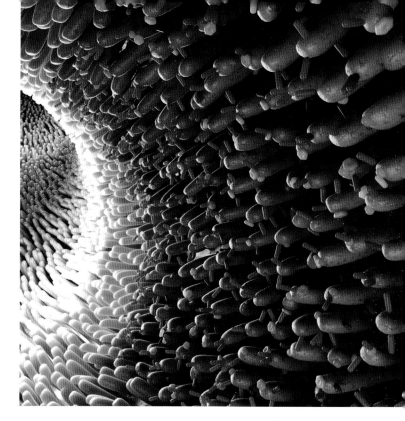

Above *Illustration of the interior of a human intestine, with numerous bacilli and cocci.*

THE GRAM TEST

All species of Bacteria are classed as either Gram-negative or Gram-positive—these two categories reflect a structural difference in their cell walls. This is tested for using a procedure called the Gram stain test, which looks at whether the bacterial cells take up a dye (crystal violet). Gram-positive bacteria absorb the dye in their cell walls and then appear stained purple under a microscope, while Gram-negative species do not.

Below *The Gram stain test involves adding a succession of chemical dyes to a bacterial sample, resulting in Gram-positive and Gram-negative species taking on different coloration.*

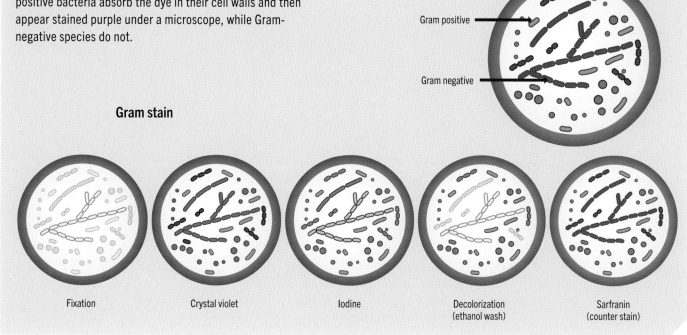

Gram positive

Gram negative

Gram stain

| Fixation | Crystal violet | Iodine | Decolorization (ethanol wash) | Sarfranin (counter stain) |

// Bacteria—cell types

The classic image we have in our minds of Bacteria is usually a mass of rod-shaped cells or bacilli, but many familiar species have a spherical form, and there are also some known as coccobacilli, which are intermediate in shape between the two. Both the Gram-negative and the Gram-positive groups contain examples of both cocci and bacilli, so these two shapes don't provide a quick and easy way to categorize bacterial types. However, all the very long spiral-shaped bacteria are Gram-negative and form a distinct taxonomic group (Spirochaetes).

Bacterial cells also show variation in how they tend to aggregate with each other. When one cell divides, the two cells may separate or remain attached to one another, and those that stay together form certain types of groups. Within the cocci, these are: diplococci (pairs of cells, which may form a capsule around themselves in some cases, for example the species that causes pneumonia); streptococci (long single chains of cells); staphylococci (large clumps or clusters of cells) and tetrads (four cells joined together on a flat plane). Other uncommon arrangements include the cube

of eight cells formed by members of the genus *Sarcina*. Bacilli cells are always joined end to end, because of the way they divide. They may be connected in pairs as diplobacilli, in long chains as streptobacilli, or be connected at their ends but stand up side by side in an arrangement known as palisade. Spirilla do not normally stay connected to each other.

The sizes of bacterial cells show a large amount of variation. This is best assessed by looking at their volume, given the variations in cell shape. One cell of an average-sized bacterium, such as the familiar rod-shaped species *Escherichia coli*, which is abundant in the human colon (and can cause illness and infection if it ends up elsewhere in our bodies), has a cell volume of between 0.4 and 3 cubic micrometers. The smallest bacteria are only 1 percent of this size (fittingly, these groups are known as ultramicrobacteria).

Opposite *The giant bacteria* Thiomargarita magnificans *(blue) shown alongside* Vibrio *(orange) to show size difference.*

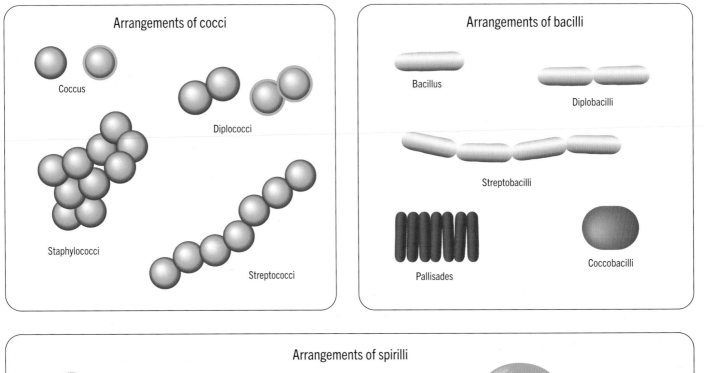

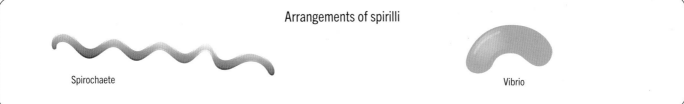

ALMOST NON-MICROSCOPIC

The biggest known species within Bacteria is the coccus species *Thiomargarita namibiensis*, with a volume of about 750 cubic micrometers—it is much larger than many eukaryotic cells and in fact is big enough to be seen with the naked eye. Almost as large are members of the rod-shaped genus *Epulopiscium*, which can reach 700 micrometers in length (exceeding the diameter of *Thiomargarita namibiensis*). Both of these bacterial giants are marine, although *Epulopiscium* lives symbiotically in the digestive tracts of fish, while *Thiomargarita namibiensis* is free-living. Some spirochaetes are up to 500 micrometers long, but their extreme slenderness (rarely more than 3 micrometers wide) means they cannot be seen without magnification.

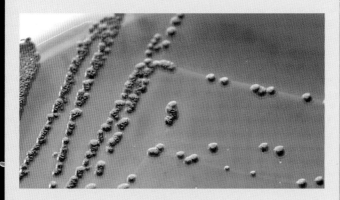

Above *Colonies of* Escherichia coli *growing on agar. Individual cells of this species are just 2 micrometers long.*

// Bacteria—metabolism

Metabolism is one of the key traits that differentiates living from non-living entities. Only a living thing (so, say a bacterial cell rather than a virus) can create its own energy stores, and gather and use material to form its own internal components. We eat food and break it down into small molecules, which we then reassemble to make our body tissues. We breathe air to provide oxygen, which we use in a chemical reaction (respiration) that breaks down glucose molecules to make a storable source of chemical energy. The glucose may come directly from broken-down food molecules if we consume sugars and starches, or can be made in our livers from other types of food molecules (gluconeogenesis). These processes are called metabolism.

In terms of getting hold of the carbon that we need to build the various molecules that form and operate within our cells, people and (nearly all) plants are, respectively, heterotrophs and autotrophs. For heterotrophs, the carbon source is organic material that we consume, whether it be plants, animals, fungi, microorganisms, or the decaying remains of all of these. For autotrophs, the carbon source is molecules of atmospheric carbon dioxide, which is broken down via the reaction of photosynthesis, and recombined with other atoms to make glucose molecules (from which other more complex compounds can be built—for example, plants build their proteins by combining glucose with nitrates taken in from soil). Most species in Bacteria are heterotrophs, but the group also includes the cyanobacteria or "blue-green algae," which are autotrophs.

The many and varied processes that take place within living cells, from molecule-building and growth to movement and reproduction, need a source of energy to drive them. This energy is derived from the sun in the case of most plants and other photosynthesizers, but for others it is derived by breaking down stored chemical energy. Organisms in the former group are known as phototrophs, while those in the latter are chemotrophs. Most autotrophs are also phototrophs, so may be known as photoautotrophs, while most heterotrophs are also chemotrophs, so may be known as chemoheterotrophs. Within the group Bacteria, all possible combinations of "chemo" and "photo," and "auto" and "hetero," are known to exist. The strange comma-shaped *Vibrio* bacteria are examples of photoheterotrophs—they use light for energy but obtain carbon from organic sources rather than carbon dioxide.

Food chain

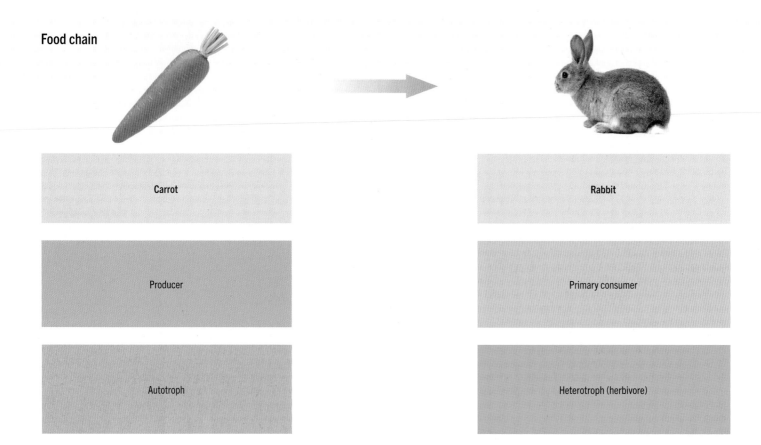

Carrot	Rabbit
Producer	Primary consumer
Autotroph	Heterotroph (herbivore)

Our usual form of cellular respiration requires oxygen (aerobic) but cells may use a different substance to play the same role (anaerobic). Some bacteria are capable of using both pathways, while others can only use one. Aerobic respiration releases water and carbon dioxide as a by-product, while anaerobic respiration produces other by-products depending on which element or compound is used in place of oxygen. For example, many gut bacteria are obligate anaerobes that use carbon dioxide rather than oxygen.

CHEMOLITHOTROPHS

A few bacteria are "rock-eaters"—able to make energy by oxidizing inorganic material, rather than carbon-containing, living (or once-living) material. This seems a near-miraculous ability to us, and it is not known among any eukaryotic organisms. However, within the group Bacteria, there is a wide array of "rock-eaters" from unrelated lineages, and some Archaea are also chemolithotrophs. Many of these organisms make use of sulfur, while others use hydrogen or nitrogen. Rarer cases include iron-metabolizers, and two species which were found in lab conditions to be able to feed on manganese.

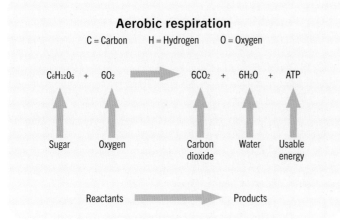

Aerobic respiration

C = Carbon   H = Hydrogen   O = Oxygen

$$C_6H_{12}O_6 \; + \; 6O_2 \; \longrightarrow \; 6CO_2 \; + \; 6H_2O \; + \; ATP$$

| Sugar | Oxygen | Carbon dioxide | Water | Usable energy |

Reactants ⟶ Products

Below A basic "food chain" shows how energy gathered by photosynthesizing plants, the producers, is passed along through various "consumers."

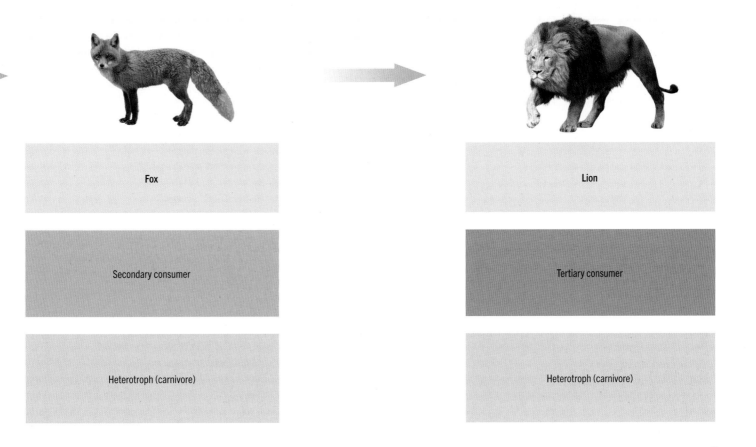

Fox	Lion
Secondary consumer	Tertiary consumer
Heterotroph (carnivore)	Heterotroph (carnivore)

// Bacteria—reproduction

We know that viruses can replicate themselves very quickly, by co-opting their host cell's enzymes and DNA/RNA copying machinery. Bacteria don't need outside help to reproduce though, and they are also capable of rapid proliferation. If you have ever explored bacterial culturing in biology classes, where a sample is placed on a growth medium such as agar jelly on a Petri dish, you will know how rapidly a large and obvious bacterial colony can develop from quite invisible beginnings.

Bacterial reproduction occurs by one cell splitting into two. This process, known as binary fission, is the same way that eukaryotic cells reproduce themselves, but happens more rapidly with simple prokaryote cells. Some Bacteria species can divide as often as every ten minutes, so after an hour, one cell has become 64. It is easy to see how a population of any such cells can grow exponentially, until imposed upon by an outside factor that limits the rate of growth. In the case of our Petri dish, the bacterial colony will eventually run out of space and food and will cease to expand (and then begin to die off). In the case of a pathogenic bacterial infection in an animal, the population growth of bacterial cells will be curtailed either by the animal's immune response successfully controlling things, or by the death of the animal host. In both cases, the bacterial colony needs to spread to somewhere new—rapid and constant reproduction is a great help to survival but is not enough on its own.

The process of binary fission occurs when the cell has grown to a larger than normal size (sometime more than double its original size). The cell then builds a duplicate copy of its DNA, and the two sets of DNA move to opposite ends of the cell. Then proteins form a circle in the center of the cell and divide the cell's cytoplasm in half. New cell membranes and walls form on either side of this split, and the single cell has now become two. The cells may remain stuck together, as in the case of species that form chains or clusters, or they may fully separate. In a few cases, the process is slightly different—sometimes one cell divides into more than two, and in some cases "daughter" cells form wholly inside the original "mother" cell, which then dies and ruptures to release the new cells.

SPORES

We associate the word "spore" with reproduction (of fungi, for example), but in the case of bacteria, a spore, or endospore to give it its full name, is something rather different. Only a few members of Bacteria can convert themselves to endospore form, whereby they divide within their cell walls, and then one half engulfs the other, the latter becoming the endospore. It contains a full set of DNA, ribosomes, and other vital materials, and can survive the destruction of the cell in which it has formed—hence endospore formation tends to be triggered by harsh conditions that threaten the cell. Endospores are inactive, require no food, and are highly resistant to extremes of temperature, drying out, exposure to many damaging chemicals and other hazards, and can reactivate or germinate when conditions become favorable again.

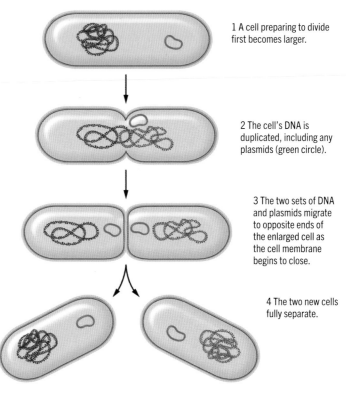

1 A cell preparing to divide first becomes larger.

2 The cell's DNA is duplicated, including any plasmids (green circle).

3 The two sets of DNA and plasmids migrate to opposite ends of the enlarged cell as the cell membrane begins to close.

4 The two new cells fully separate.

Right *The stages of bacterial cell division.*

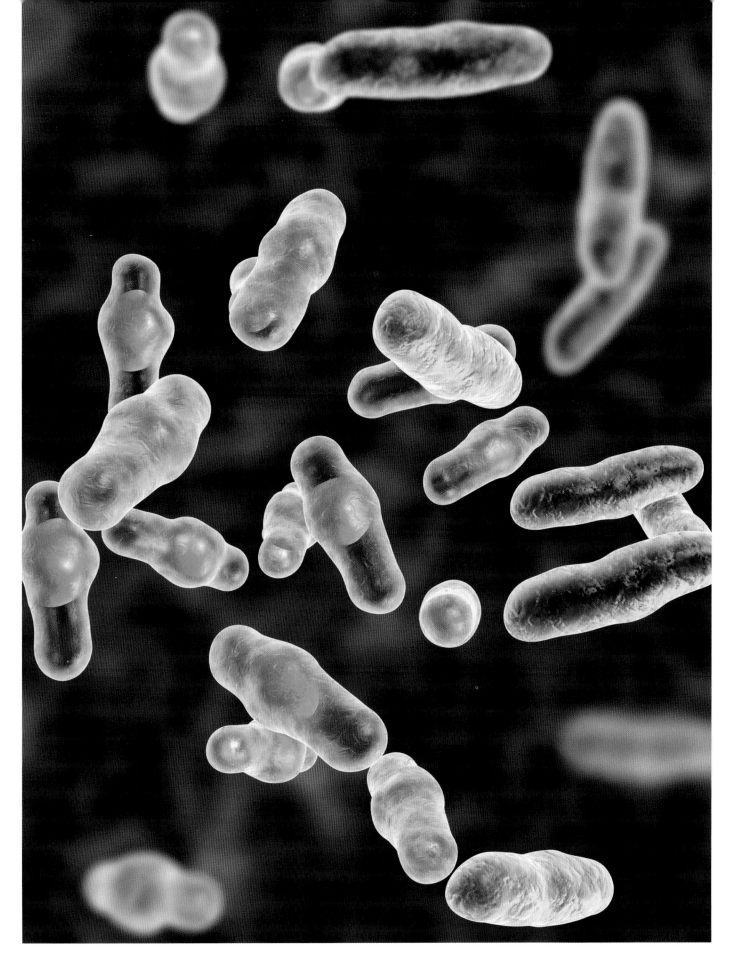

Above *This illustration of bacilli shows some in the process of forming endospores.*

// Bacteria—abundance and diversity

As we have seen, the domain Bacteria makes up a very substantial proportion of all life on Earth, not only in abundance of individuals, but in terms of biomass (the collective weight of all individuals). Given the tiny size of each cell, it's very clear that Bacteria exist on Earth in the sorts of numbers that are difficult for us to wrap our heads around. Collectively, there are an estimated $4-6 \times 10^{30}$ prokaryotic cells in the world, the majority of them Bacteria. They occur throughout the biosphere—on and in water and land of all kinds, and on and in the bodies of eukaryotes like us.

So we know they are everywhere, and we know that they come in different types, but how many different types? Working out a way to classify such minuscule organisms objectively and scientifically is no easy task. Microscopy, much advanced though it is today, can only show us rather crude differences between different types of Bacteria. If we were to attempt to classify them by appearance, the criteria we could use are rather limited. Cell shape and the way the cells tend to arrange themselves would give us a few categories, while factors like size, the presence or absence of pili, flagellae, or other visible structures would provide a few

more. Other factors such as their preferred habitat and food, and the Gram staining test, allow for further distinctions.

It was not until the advent of gene sequencing techniques, first devised in the late 20th century, that we had a tool that allowed us to properly explore the enormous diversity of bacterial life on Earth. Through genetic study we can pin down differences between different strains of bacteria on the molecular level. Rapid progress has occurred since then, but this is still a young science and we are a long way away from establishing a system of classification for prokaryotes that is as advanced as that which we use for eukaryotes (which is itself still far from complete). For the domain Bacteria, biologists estimate that about 1,300 phyla exist. However, the most widely used formal databases of prokaryote names accept fewer than 100 phyla so far, with many more proposed phyla awaiting assessment (and many more presumed to be undiscovered).

Below *Transcribing the full genome of even a simple organism such as a bacterium is a lengthy and painstaking task.*

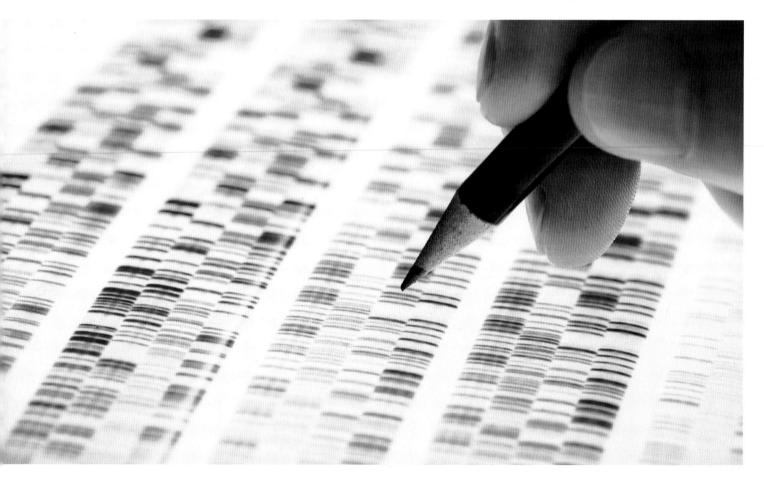

BACTERIAL SPECIES

Several of the bacteria that have an intimate relationship with our own species have been gene-sequenced and go by a scientific name that follows the same rules as those we use for eukaryotes—a genus name and a species name. For example, our scientific name is *Homo sapiens*, and our colons are full of the bacterial species *Escherichia coli*. However, this may not be the best way to define bacterial species. The concept of a species traditionally included the idea that, if individuals from a population could breed freely together, then they were of the same species. But this is irrelevant to asexually reproducing organisms such as bacteria. Then there is the fact that bacteria exchange genetic material with each other via a non-reproductive route (horizontal gene transfer, as opposed to vertical gene transfer from parent to offspring)—this is a key part of how they evolve and gain genetic diversity. A completely different approach to bacterial classification may therefore be more suitable, but this is not yet agreed upon.

Above *Evolutionary pathways of the main lineages within Bacteria, as indicated by genetic comparison. A tree diagram like this is very useful to show how lineages arose and diverged, but cannot capture the full picture as it does not account for genes transferred between different lineages, rather than passed on from parent to offspring. About 8 percent of our own DNA consists of genes that originated from retroviruses, which infected our ancestors many millennia ago.*

// Bacteria—cyanobacteria and photosynthesis

One of the better-known and -studied phyla within Bacteria is Cyanophyta, or the cyanobacteria (often commonly known as "blue-green algae," but misleadingly so as true algae are actually eukaryotes). These organisms are important to the history of life on Earth and of eukaryote evolution because they were the original photosynthesizers, using sunlight to drive their energy-building metabolism, and producing oxygen as a by-product. Had this chemical innovation not come about, 3.5 billion years ago, Earth would not have developed the oxygen-rich atmosphere that it has now, and our own evolution would not have happened.

Today, cyanobacteria are still very widespread and abundant, and are a crucial, foundational part of our planet's ecology as primary producers, obtaining their energy from light rather than by consuming other materials. Marine cyanobacteria are responsible for about 50 percent of the primary energy production on Earth, and their role in the carbon cycle is also critical. As well as living freely in water (both marine and fresh), cyanobacteria occur in the soil, are carried high into the air on sea spray, and also exist in intimate symbiotic association with other organisms. Some water-dwelling cyanobacteria form large and very visible colonies, aggregating into sheets or delicate filaments.

Their photosynthesis relies on certain structures not found in other Bacteria lineages. The pigments that are chemically changed by exposure to sunlight during photosynthesis are found in structures called phycobilisomes, which are held within an interior membrane (the thylakoid membrane). Cyanobacterial cells also contain structures called carboxysomes, which accumulate the carbon dioxide required for photosynthesis. By night, cyanobacteria cannot photosynthesize and this is when they release oxygen. Although they are responsible for creating our life-sustaining atmosphere, they also produce a range of highly toxic chemicals, which can cause illness in humans and animals. You may have seen signs warning of "algal blooms" around lakes—this indicates that conditions have caused cyanobacterial populations to increase dramatically, which means that concentrations of their toxins become high enough to be dangerous and bathing should be avoided.

Below *Although it is truly life-giving, cyanobacteria can also produce dangerous toxins.*

Below *Lichens come in a variety of forms, and a rich lichen community is indicative of good air quality.*

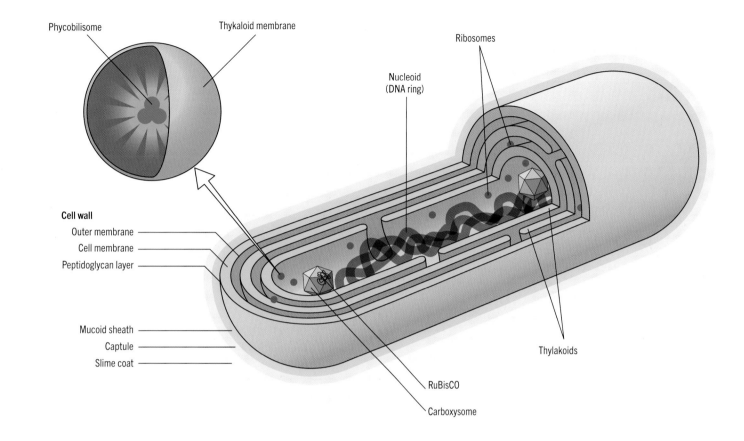

Phycobilisome

Thykaloid membrane

Ribosomes

Nucleoid
(DNA ring)

Cell wall

Outer membrane

Cell membrane

Peptidoglycan layer

Mucoid sheath

Captule

Slime coat

Thylakoids

RuBisCO

Carboxysome

Above *Diagram of a single plant cell.*

LICHENS AND MORE

Lichens are familiar components of many land habitats. You'll see them forming patches on old stonework and draping the branches of trees—some look like a simple "painted-on" encrustation while others have a more elaborate form, resembling tiny fleshy leaves or fine tendrils. All lichens are composites, comprising a photosynthesizing organism (a cyanobacteria or eukaryotic true alga—sometimes both) entwined with the mycelia of one or more species of fungus. The photosynthesizer supplies energy while the fungus captures and holds water and collects nutrients from the substrate the lichen grows upon. A handful of animals, primarily certain sponges, corals, and sea anemones, also have symbiotic relationships with photosynthesizing cyanobacteria and algae. The link between plants and cyanobacteria, though, is more ancient, ubiquitous, and fundamental.

// Bacteria's role in nutrient cycles

The oxygen, carbon, hydrogen, and nitrogen that all living things need to build their cells move around our world in a series of cyclical processes. A simple and familiar example concerns oxygen—plants release it when they photosynthesize, and we breathe it in. With carbon dioxide (a compound of carbon and oxygen) the process is reversed—we exhale it and plants take it in. But there's more to it than that—the plants also need oxygen, and we also need carbon. These elements, along with hydrogen and nitrogen plus smaller amounts of many others, are constantly taken up and released by organisms of all kinds, including the unthinkably vast amounts of organisms that form the domain Bacteria.

Below *The nitrogen cycle involves several types of decomposing organisms (primarily bacteria), breaking down organic material by stages, eventually producing nitrates that plants can utilize.*

Above *Droppings might just be waste to the animal that dropped them, but a huge community of microbes and other organisms is adapted to reclaim the nitrogen and other nutrients they contain.*

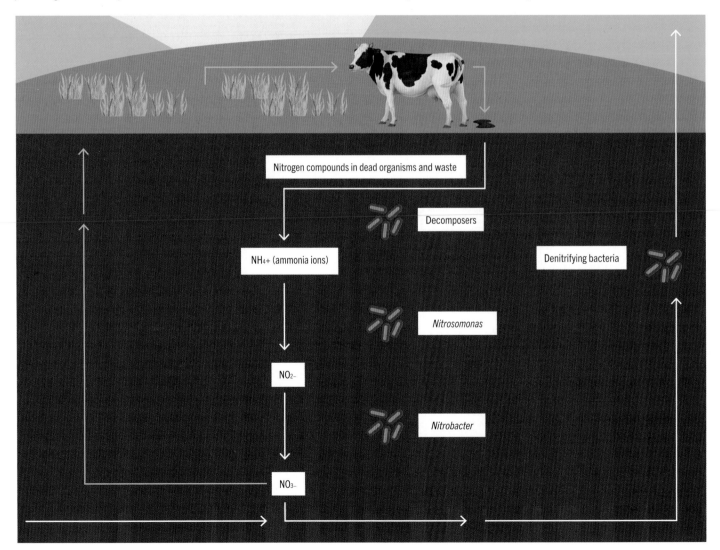

Nitrogen compounds in dead organisms and waste

Decomposers

NH_4+ (ammonia ions)

Denitrifying bacteria

Nitrosomonas

NO_2-

Nitrobacter

NO_3-

Above *Root nodules on a soya plant—home to nitrogen-fixing bacteria.*

Bacteria that live in the soil play an important role in recycling nutrients that fall to the ground, in the form of animal droppings, autumn leaves, and dead organisms of all kinds. They break large organic molecules down into small, water-soluble molecules that plants can take up in their roots. They can also capture the nutrients in fertilizers and convert them into different forms, which plants can easily use. Some bacteria can capture ("fix") atmospheric pure nitrogen and make it available for plants as well as other soil microorganisms—certain plants actually have populations of nitrogen-fixing bacteria residing in nodules that form on their roots. Cycling of the element phosphorus between inorganic and biological systems is particularly reliant on the actions of bacteria, as plants cannot extract it directly from the rock in which it naturally occurs. The sparser the amount of phosphorus in an environment, the more likely it is that there will be bacterial species that are efficient at cycling it. Bacteria are also closely involved in the water cycle—photosynthesizers use it and those that use aerobic respiration release it. Bacterial cells in the atmosphere even encourage clouds to form, as water vapor condenses around them.

DIAZOTROPHS

The name diazotroph describes the numerous types of Bacteria that can convert or fix atmospheric nitrogen into a different form, which is useable by other organisms. Some diazotrophs live freely in the soil while others are associated with plants. Some are capable of photosynthesis as well as nitrogen fixation, while many use no oxygen in their own metabolism (they are anaerobic organisms) and can only operate in very low-oxygen environments. Atmospheric nitrogen takes the form N_2—a pair of atoms joined into a very stable molecule, which we take into and release out of our bodies with every breath without affecting its structure at all. Because we and other organisms can't readily use atmospheric nitrogen for ourselves, we must take it on in other forms—in the plant and animal matter that we eat. Plants such as grasses take in nitrogen in the form of nitrates, created by diazotrophs, from the soil. Animals obtain nitrogen "second-hand," by consuming plants. All living things need to obtain nitrogen from somewhere, because this element is a key component of proteins, the many forms of which are essential to build cells and carry out a host of metabolic processes.

// Bacteria as commensal organisms

Many of us are increasingly aware that our bodies are not entirely, or even mostly, human in nature. In terms of cell count, we are more Bacteria (and a bit of Archaea) than we are animal. These freeloading organisms are described as commensal, meaning that they live in association with us, but in some cases the associations are also mutualistic—beneficial to us and them. Many of these bacteria dwell in our intestinal tract, where they carry out a range of helpful functions that we could not live without.

Our intestinal microbiome is highly individualistic and influenced by many factors, from age and diet to our genes and whether we were born before or after our due date. Any given human gut contains in the region of 300–500 different species, and their activities range from digesting dietary fiber to improving our immune response and helping with vitamin absorption. There is increasing evidence of a strong link between mental health and the health of the gut biome in mammals, although how this association works is still a matter of mystery.

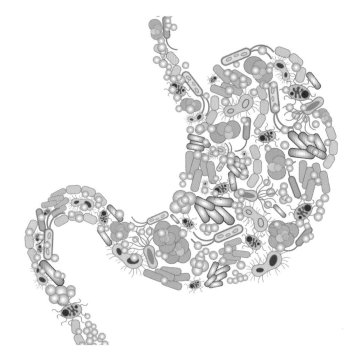

Above *Our stomach is not just an organ—it is also a habitat for a whole ecosystem of Bacteria and Archaea.*

Left *Commensal bacteria are responsible for the stunning bioluminescence of this bobtail squid.*

Above *Breaking down a meal of tough grass is a team effort. The cud chewed by cattle has already been subjected to the attention of fermenting bacteria within the animal's rumen.*

The resident bacteria in our noses and mouths are also part of our immune response, able to help protect us from pathogens that may enter via these routes. The bacteria *Staphylococcus hominis*, which occurs abundantly on our skin, is responsible for generating body odor in our sweaty armpits—a bad thing by today's standards, perhaps, but our individual scents may have been important in individual recognition and even attraction in our ancestral past.

Bacteria live in commensal and mutualistic relationships with many other organisms, too. We have already seen how some plants host populations of nitrogen-fixing bacteria in their roots. Some marine animals have the ability to glow in the underwater darkness thanks to bioluminescent bacteria living on their bodies—the anglerfish uses its glowing "lure" to attract prey, while some squid use bioluminescence as a form of camouflage, eliminating their dark silhouette if seen from below against a moonlit sky. The blue-ringed octopus, one of the most feared venomous sea animals with a bite that can quickly kill a human, derives its venom from a community of *Vibrio* bacteria that live in its salivary glands.

RUMINATIONS

Bacteria are essential in our guts but play an even more elaborate role in the digestive tract of cows and other multi-stomached grazing mammals (ruminants). Grass is a tough substance to break down and extract nutrition from, and those relatively few animals that eat it have several tricks to help them. Cows chew the cud, regurgitating mouthfuls of partly chewed grass back into their mouths from the rumen (the "first stomach") for further mastication, but also have a huge population of fermenting bacteria within the rumen and reticulum (the "second stomach"), which carry out chemical breakdown of the fibrous cellulose in the grass. Other fermenting microorganisms including fungi, Archaea, and some protozoa are also present in the rumen's microbiome. Only after much activity by these microorganisms can the rest of the digestive system actually absorb any nutrients from the consumed grass.

// Bacteria as pathogens

Like viruses, bacteria can be pathogenic, attacking and causing disease in other organisms, and often spreading between them. Sometimes the same bacterial species may be harmless and even beneficial in some contexts but harmful in others. For example, *Escherichia coli* lives in abundance in our colons and causes no problems, but if we inadvertently ingest *E. coli* of the strain O157:H7, it causes a very unpleasant stomach upset and, in a few cases, life-threatening kidney failure. Harmless bacteria that live on our skin may cause an infection if they enter an open wound.

Pathogenic bacteria cause harm in a number of ways. They may cause direct damage or destruction to the host cell that they bind to or enter, or they may release toxic substances that harm or kill the cells around. This release of toxins may occur continually, or when the bacterial cell is destroyed by the host's immune system. Sometimes it is not the bacteria itself but an extreme immune response from the host that does the damage—sepsis is an example of this.

Diseases caused by bacteria are important factors in the lives and deaths of many other organisms. Many animals have a range of natural behaviors that help protect them from catching bacterial infections, ranging from keeping wounds clean and avoiding drinking water from certain sources, to spreading out as best they can when conditions are overcrowded. Plants are also more susceptible to infectious disease when they grow as a monoculture, as bacterial infections can spread between individuals growing close together through splashing rain water, for example. The outer surfaces of animals and plants are adapted to be resistant to bacterial invasion, too. However, bacteria are so abundant and so tiny that it is difficult to keep them out all the time, and organisms often have to rely on their immune systems to deal with an infection. The outcome then is highly variable. An infection that would go unnoticed by an individual with strong immunity may be very serious for another whose immune system is weak or compromised.

Below *Petting zoos are delightful places for young and old, but it's very important to maintain good hygiene or you could bring home more than you bargained for.*

Please be aware of the risks of
E.coli 0157
Through contact with animals.
Remember to wash your hands
Handwashing facilities are
available in the shop

FROM DISCOVERY TO DETECTION

The first scientist to observe examples of bacteria and other microorganisms was Antoni van Leeuwenhoek, a Belgian merchant and builder of microscopes. Discovering in the 17th century that these creatures lived not only in pond water but also in his own mouth, he was unenthralled and took to gargling with vinegar to try to kill them off. Another 200 years elapsed, though, before microorganisms were recognized as possible disease-causing agents in humans. The work of 19th-century doctors and scientists, such as Louis Pasteur and Robert Koch, provided the foundation for our understanding of how microorganisms cause infections, and ways that we can combat them (including the first vaccines).

Above *Antoni van Leeuwenhoek.*

Right *These discolored soya leaves are affected by infection with the bacteria* Pseudomonas syringae *pv.* glycinea.

// Bacterial diseases of humans

Back in the 1800s, 1900s and early 20th century, humans died in great numbers from a disease known as "consumption." Another name for it was "the white plague," because of the characteristic paler skin that sufferers developed. This disease today is known as tuberculosis, and it was found in the late 19th century to be caused by an infective agent, spread mainly through coughs and sneezes, which was recognized as a bacterium and named *Mycobacterium tuberculosis*. This sparked a program of public health efforts, such as campaigning to stop people from spitting in public, but it was not until the mid-20th century that effective antibiotic treatment became available.

Below *Patients and their nurses on a tuberculosis ward in Lyon, France, taken between 1914 and 1920.*

Tuberculosis is one of several bacterial diseases that have caused widespread misery to our species over the years. Others include tetanus (caused by *Clostridium tetani*), cholera (*Vibrio cholerae*), diphtheria (*Corynebacterium diphtheriae*, leprosy (*Mycobacterium leprae*), and the bubonic plague (*Yersinia pestis*). A variety of other Bacteria species can cause throat and ear infections, food poisoning, urinary tract infections, and other milder but still highly unpleasant illnesses. They spread from person to person in various ways, including droplets released in the air from coughs and sneezes, direct skin-to-skin or sexual contact, or via contaminated food and drinking water. However, out of the thousands of species of Bacteria that we have described, fewer than 100 are known to cause disease in humans.

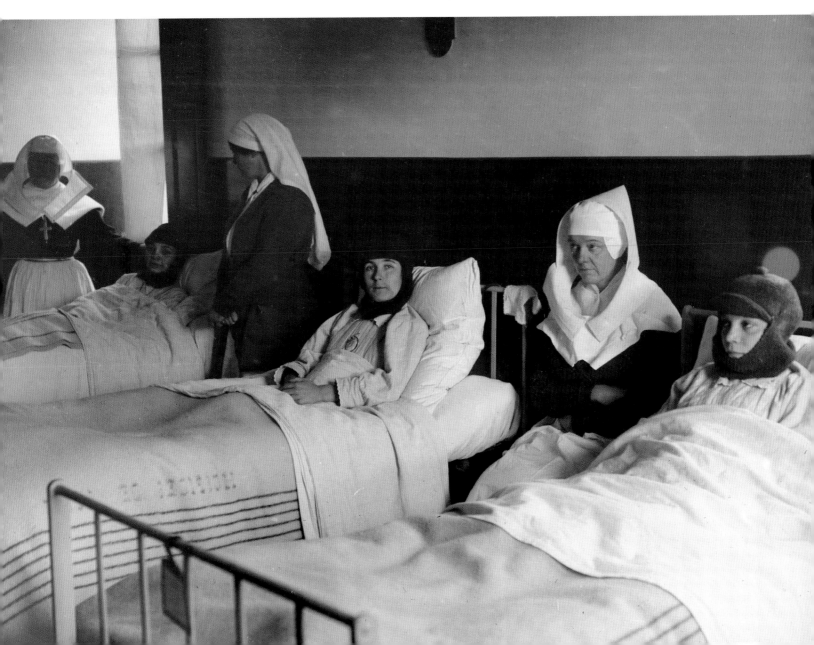

Humans are social by nature and in some parts of the world we spend much of our time in very crowded conditions. This makes us much more susceptible to bacterial diseases than more solitary animals. Keeping ourselves from catching or spreading bacterial diseases depends on as many people as possible adopting good personal hygiene and good habits—hand-washing, using a handkerchief when coughing or sneezing, drinking clean water, and sticking with good food hygiene protocols. We also have medical options if we do develop a serious infection (see page 66), and vaccines have been developed (although their availability varies a lot by location) for some of the deadliest bacterial diseases.

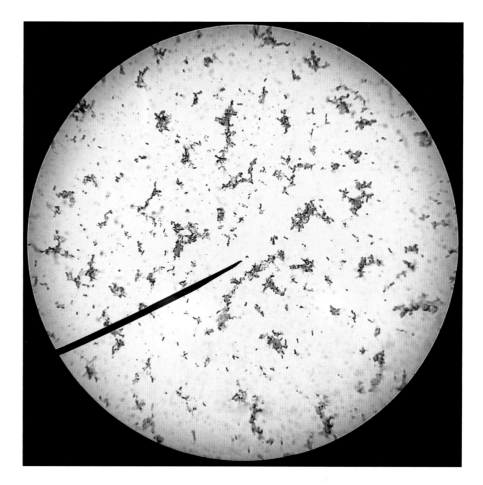

Right *Sample of* Vibrio cholerae, *the cause of cholera*

PROGRESSION AND PROGNOSIS

You can catch tuberculosis, still a leading cause of death in adult humans worldwide, when you inhale the bacterial cells carried in droplets that are breathed, coughed or sneezed out by an infected person. Whether the bacteria then takes hold in your own lungs depends how large a dose of bacteria you inhaled, and how effective your own immune response is. This response will kick in straight away, with macrophage cells destroying bacterial cells, but any bacteria that survive will reproduce rapidly and your immune response may be overwhelmed. The bacterial cells move through your lungs and may invade other tissues via your lymphatic system. Meanwhile, your adaptive immune response (see page 28) begins to get to work, destroying the bacterial cells with greater efficiency, but if it too is overwhelmed, full-blown tuberculosis will develop as the bacterial cells proliferate and cause extensive damage. By this point, you will probably need medical intervention if you are to recover.

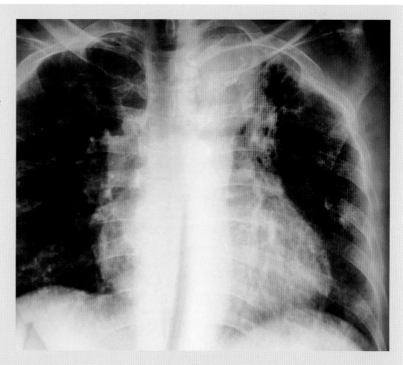

Above *A lung x-ray reveals the damage done by a tuberculosis infection.*

// Antibiotic agents

Bacteria are living things, and so they can be killed. When pathogenic bacterial cells enter your body, a lot of them will be killed by the cells in your immune system, but in immunocompromised people and those who have an extensive infection, this may not be enough. The answer that medical science provides in these cases is a choice of drugs that will assist your immune system by (hopefully) killing the bacterial cells. However, not all antibiotics directly kill the bacterial cells. Some of them interrupt certain vital processes and prevent the cells from reproducing instead, meaning that the infection will not spread rapidly and so your immune system has many fewer bacterial cells to target and destroy.

We use antibiotic agents in other ways besides medicinally—for example in hand sanitizers and other kinds of disinfectant. These surface cleaners, widely used in hospitals and food preparation areas, are able to target a large range of species, and the same goes for some medicinal antibiotics—these are known as broad-spectrum. Narrow-spectrum antibiotics target fewer species but usually with more effectiveness at eliminating those particular species.

The advent of antibiotics through the middle of the 20th century brought about a huge improvement in human health

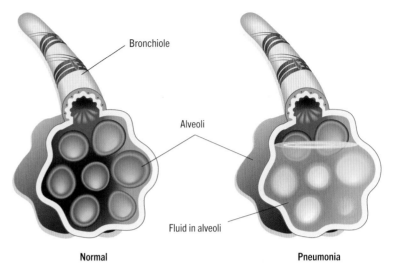

Normal   **Pneumonia**

Bronchiole

Alveoli

Fluid in alveoli

Above *Pneumonia, usually caused by* Streptococcus pneumoniae *infection, causes the alveoli in victims' lungs to fill with fluid, and may require intravenous antibiotic treatment.*

Below *Regularly and thoroughly cleaning surfaces with bacteria-killing agents is one of the most important measures we can take to stop the spread of disease.*

and life expectancy, as hitherto deadly diseases became highly treatable. The use of antibiotic disinfectants reduced the incidence of infections in the first place, so much so that complex and lengthy open surgeries—which would previously have carried a huge risk of large-scale infection—became much safer.

Right *One of the first antibiotic medications to be licensed, penicillin is still widely used today.*

ANTIBIOTIC RESISTANCE

Of the many possible threats to our own species' survival, antibiotic resistance is one of the more immediately concerning. As we have seen, bacterial cells can replicate very rapidly, and in the process of replication, a few random DNA mutations occur that mean that offspring cells may have slight genetic differences to parent cells. If we are considering an active pathogenic infection that is being treated medically, and a cell happens to acquire a mutation that makes it resistant to the antibiotic being used, that cell will certainly survive and so will its descendants. The infection can then quickly "bounce back" as the resistant strain proliferates freely. Another antibiotic must now be tried. This is how natural selection works, and it is of grave concern for us because we do not have the ability to produce an endless supply of different antibiotic agents. This is why it is important that antibiotic use is regulated, to slow down the inevitable development of antibiotic resistance in more and more bacterial strains.

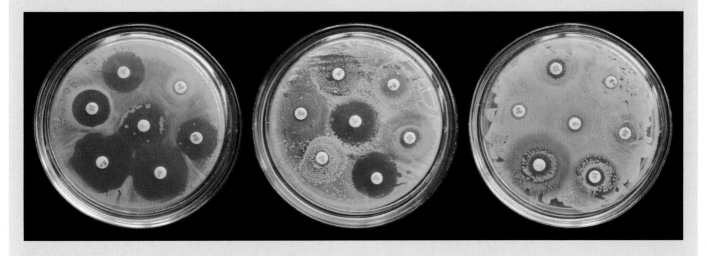

Above *How antibiotic resistance develops over time, with the agent progressively less effective at inhibiting bacterial spread.*

// Bacteria—uses

Our biological associations with the bacteria that live inside us date back to long before we became the species we are today. Like other animals, we rely on them as symbiotic organisms, just as much as we are vulnerable to them as pathogens, and their roles as primary producers and nutrient recyclers are vital to the continued existence of life on Earth. Bacteria have a host of other potential uses specifically for our human needs, too, which we are just beginning to explore and embrace.

Like other cells, bacteria have the ability to build proteins and other compounds, and we can manipulate the kinds of compounds they make to suit our own needs. In 1978, a biotech company first used genetically modified *E. coli* to produce insulin, the hormone used by diabetics to manage their blood sugar levels. The technique concerned, recombinant DNA technology, involves implanting a modified version of the bacterial genetic code into host cells, and then culturing a large population from those descendants of the original host cell that do indeed synthesize the desired compound.

Bacteria have uses as consumers as well as producers of chemicals. In 1975, a strain of the species *Paenarthrobacter ureafaciens* was found in Japan that was able to digest the synthetic material nylon. It had evolved in ponds that received run-off water from a nearby nylon factory, and when, by chance, a mutant form appeared that was able to digest nylon, that mutant's descendants thrived in their particular habitat. Now, biotechnologists are working to develop a range of bacterial strains that can digest different kinds of plastics, as a way to combat plastic pollution. Bacteria can also be used to break down pesticides and other pollutants, releasing harmless by-products.

Another important use of bacteria is as model organisms in lab settings, for the study of various cellular processes, and genetic inheritance. Because they are easily cultured and have a rapid generation time, they are ideal for studying how genetic mutation and selective pressures drive evolution and lead to the emergence of new traits.

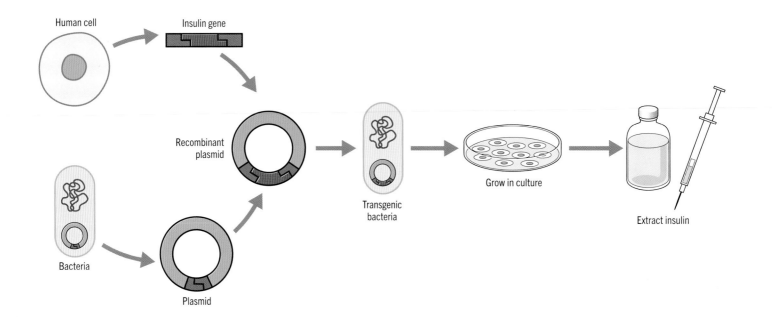

Above *How bacteria are used to make insulin.*

FERMENTATION

As we have seen, some bacteria carry out their energy-building metabolic process (respiration) in the absence of oxygen, and this is known as fermentation. While aerobic bacteria produce carbon dioxide and water as by-products, the anaerobes produce different by-products, and some of these may be useful to us. One example is alcohol, which has many industrial uses as well as being an ingredient in intoxicating drinks. It is formed when certain bacteria (and some other organisms such as yeasts) break down sugars anaerobically. Some other anaerobic bacteria generate lactic acid, which is responsible for the tangy taste of fermented foods such as kefir, yoghurt, and kimchi. The live bacteria communities in these foods are believed to help support our own gut health when we consume them.

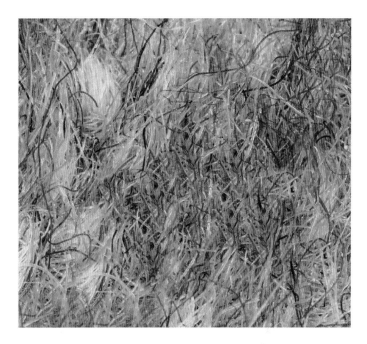

Below *Kimchi, a dish of spicy fermented vegetables such as cabbage and carrot, is a healthy and increasingly popular choice.*

Above *Nylon fibers make for a very hard-wearing carpet, but some bacteria have evolved the ability to use this material as food.*

// Endosymbiosis

We humans and other non-microscopic organisms are eukaryotes—we could be defined as "complex life." Even microscopic, single-celled eukaryotes are much more complex than any prokaryotic cell. Bacteria and Archaea existed and thrived for well over a million years before eukaryotic life first appeared. There was nothing inevitable about the advent of eukaryotic life—but once it did appear, the course of evolution took a new direction, and it came into being through associations between the various prokaryotes that already existed.

It has long been recognized that one of the organelle types found in eukaryotic cells, the mitochondria, are rather different to the others. They have double membranes rather than single, and contain their own ribosomes and, most tellingly, their own DNA (which they pass down their own lineage as they divide, independently of the surrounding cell). These factors strongly suggest that they have a different origin to the cell as a whole. However, they are very much a key functional part of the cell, being responsible for respiration—building the molecules of adenosine triphosphate (ATP) that is the cell's store of chemical energy, to be broken down whenever energy is required. In appearance, structure and function, they closely resemble free-living members of Bacteria, in particular the group Alphaproteobacteria.

Plant cells contain mitochondria, and also chloroplasts, which are the sites of photosynthesis. Like mitochondria, chloroplasts hold their own DNA, and they too resemble members of Bacteria, specifically the cyanobacteria that were the Earth's first photosynthesizing organisms. The extreme similarity of mitochondria and chloroplasts to prokaryotic Bacteria cells led to the theory that eukaryotes arose through an archaean cell (which may have already formed a nucleus) engulfing bacterial cells, and the two organisms effectively becoming a single functional unit over time—endosymbiosis. This is the most widely accepted modern theory to explain the origins of eukaryotes. The nature of modern-day Alphaproteobacteria adds further support to the theory. This group is a diverse assemblage of gram-negative species that include some pathogens and some parasites, but also some that live as endosymbionts within the cells of modern animals.

Below *The theoretical stages of endosymbiosis which led to the evolution of eukaryotes.*

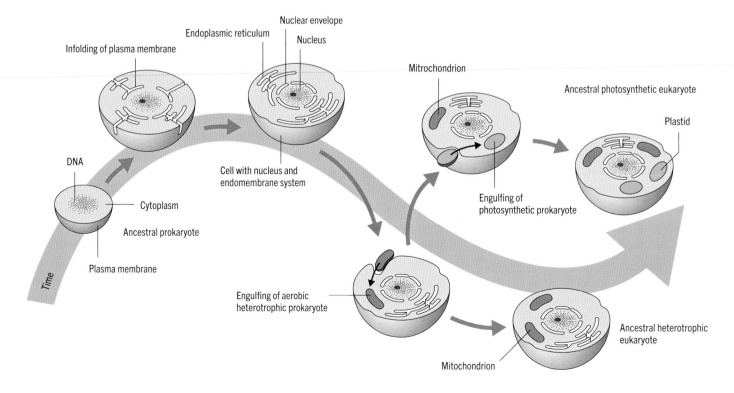

Mitochondria and chloroplasts today are different to their free-living Bacteria relatives, having adapted to their new circumstances over millions of years. Likewise, eukaryotic cells differ greatly from any modern members of Archaea, having evolved into a larger and more complex form with a range of additional organelles as well as the cell nucleus. However, it is highly likely that we and all other eukaryotes arose from the most intimate union of the two great prokaryotic domains of life—a fortuitous event that may have only happened a handful of times over the long history of prokaryotic life.

Below *The chloroplasts in plant cells are numerous and highly visible.*

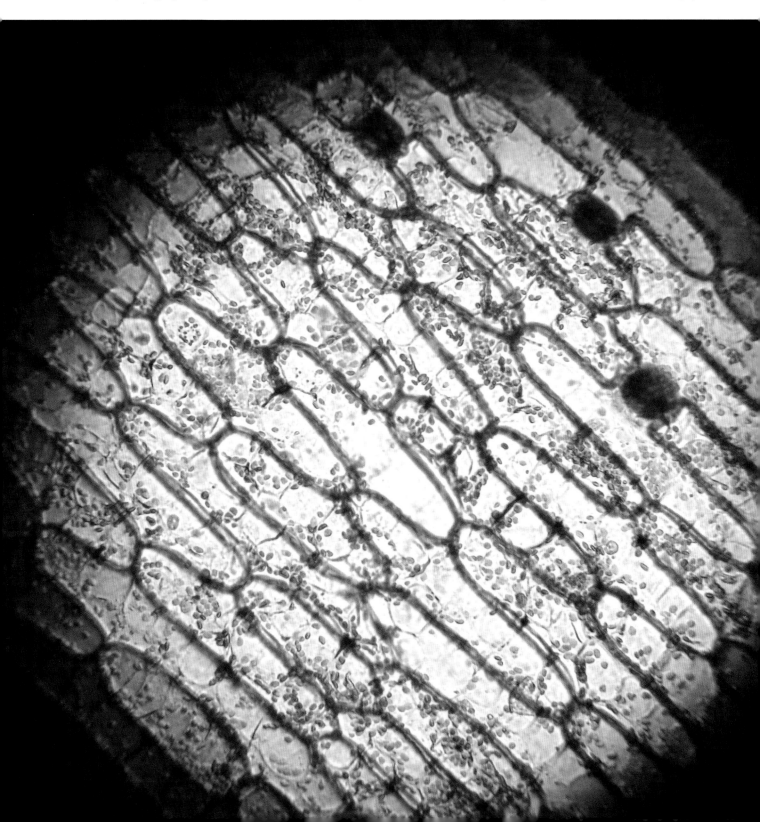

EUKARYOTES

The single-celled organisms we may see swimming across a microscope slide are large and complicated in structure and behavior compared to prokaryotes. They propel themselves with efficiency, capture food particles and avoid hazards on their way, some can capture sunlight to build energy stores, and a few can even "see" where they are going. So much functionality packed into a single cell is impressive, and comes courtesy of the organelles that are a defining feature of all eukaryotes. As their name suggests, organelles are structures within cells that behave like miniaturized versions of the organs in our bodies. Two key types of organelles, the energy-generating mitochondria and the photosynthesizing plastids, descend from free-living prokaryotes, which became functional parts of larger cells after being engulfed by them in a pivotal evolutionary process called endosymbiosis.

Mitochondria are functional organelles within almost all eukaryotic cells, but they were once free-living prokaryotes.

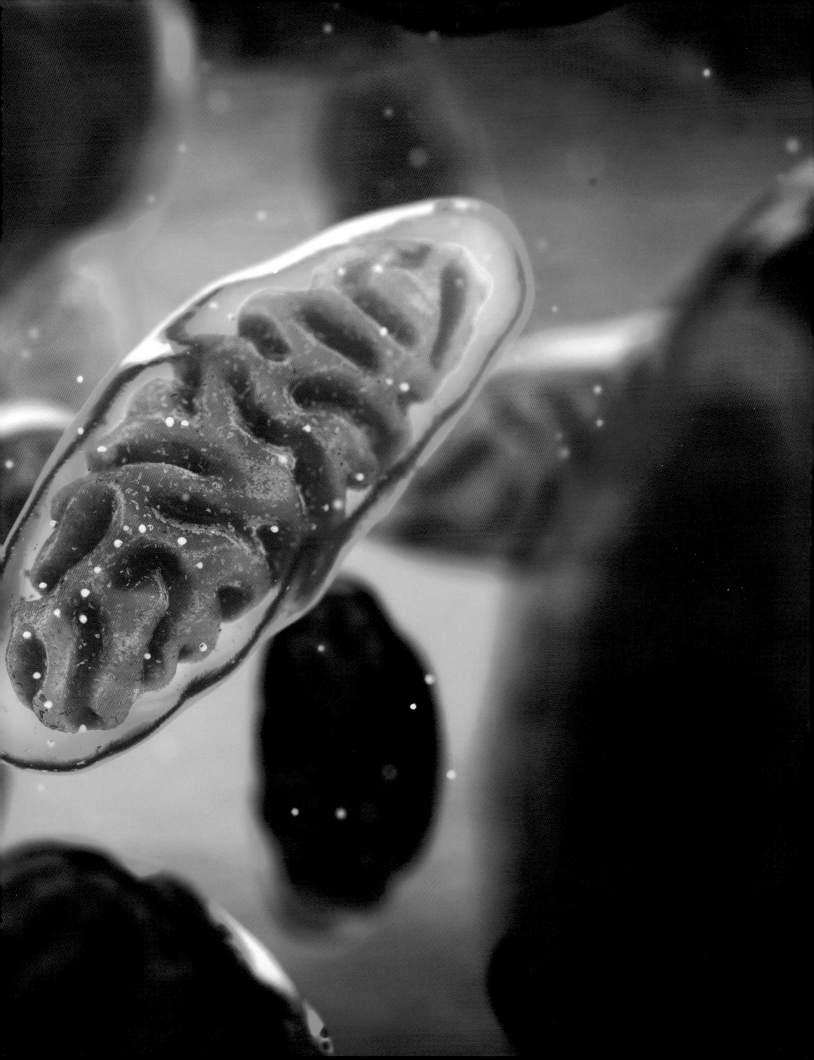

// What defines a eukaryotic cell?

Thanks to the fortuitous combining of prokaryotes, at least 2.7 billion years ago, the way was paved for a new array of life forms to evolve on our world. Eukaryotes are the descendants of Archaean cells and hold mitochondria and in some cases chloroplasts (both the descendants of bacterial cells) within them, but much more besides. A typical eukaryotic cell is much larger than a prokaryote and contains numerous different kinds of organelles in addition to the ones that are of bacterial origin.

Below *A multicellular eukaryote, in this case, a human blastocyst (a fertilized egg that has gone through its first few rounds of cell division).*

The world of eukaryotes today includes animals, plants, and fungi—multicellular living things that we might describe as "higher" or "complex" life, and all living things that are truly non-microscopic (as individual organisms) and can be studied meaningfully with our naked eye are eukaryotes. We don't need to peer into their cells to know that a tree, elephant, or mushroom is a eukaryote. However, the domain Eukaryota also includes a great variety of single-celled, microscopic organisms, which we can only watch and study under magnification and that interact with prokaryotic life in very different ways to macroscopic organisms.

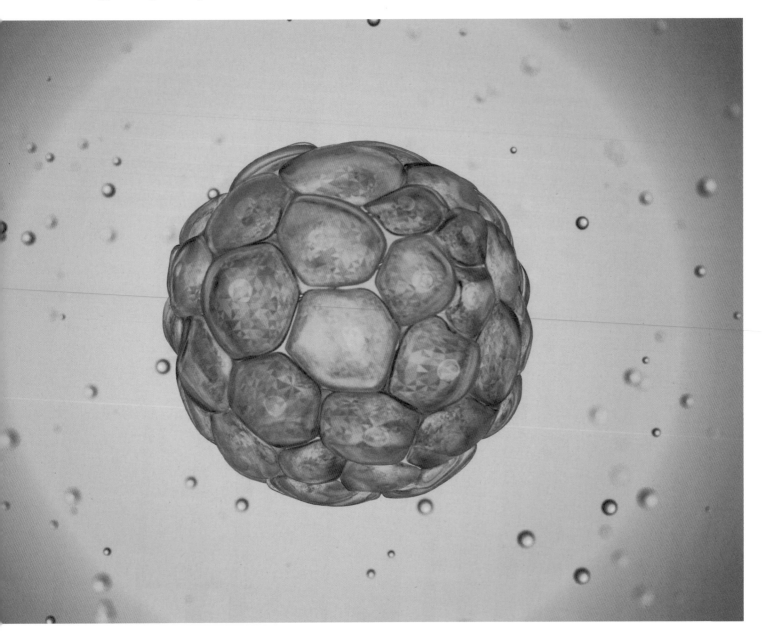

Below *Differences between prokaryotic and eukaryotic cells.*

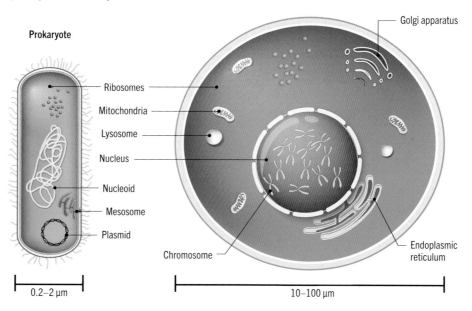

Prokaryote

Eukaryote

Ribosomes
Mitochondria
Lysosome
Nucleus
Nucleoid
Mesosome
Plasmid

Golgi apparatus
Chromosome
Endoplasmic reticulum

0.2–2 µm

10–100 µm

The most important distinction between a eukaryotic and prokaryotic cell is that the eukaryote keeps its DNA within a distinct membrane-bound organelle—the nucleus. The eukaryotic cell also contains a range of other organelles, which we explore later in this chapter. Each organelle has a specific function, although it often works closely with other organelles, and collectively they can be likened to the system of organs within an animal's body. With all this extra content, it is no surprise that the average eukaryotic cell is 20 times larger than the average prokaryotic cell.

To understand fully how a standard eukaryotic cell is built and how it works, the "best" examples to examine are the single-celled organisms within the group, for which one cell carries out all functions necessary for life.

In complex multicellular organisms, such as us, cells form distinct tissues with specific functions, for which they must differentiate into specialized types. They are still self-regulating to at least some extent but are not self-sufficient and rely on support of various kinds from other cells around them, and other cell types elsewhere in the organism's body. These specialized cell types may contain additional structures (such as the contractile protein fibers in muscle cells) or lose other structures (such as the loss of the nucleus in red blood cells). However, mature specialized cells do arise from stem cells that have a more generalized structure, and they only fully develop their special traits after several rounds of cell division.

Prokaryotic cells	Eukaryotic cells
Prokaryotic cells don't possess nucleus and membrane-bound organelles.	Eukaryotic cells possess membrane-bound organelles including the nucleus
Normally 0.2 to 2 µm in diameter.	Normally 10 to 100 µm in diameter.
Have no true nucleus, no nuclear membranes and nucleoli.	Consist of a true nucleus with nuclear membranes and nucleoli.
Consist of a single, circular DNA molecule in the nucleoid.	Consist of multiple, linear chromosomes in the nucleus.
Membrane-bound organelles are absent.	Membrane-bound organelles are present.
Flagella are made up of two proteins.	Some cells without cell wall contain flagella.
Mostly made up of peptidoglycans.	Made up of cellulose, chitin, and pectin.
Carbohydrates and sterols are not found in the plasma membrane.	Carbohydrates and sterols serve as receptors on the plasma membrane.
Small in size.	Large in size.
Cell division occurs through binary fission.	Cell division takes place through mitosis.

// Forms of eukaryotic cells

Throughout the living world, there is vast variety in the general form of eukaryotic cells. While Bacteria and Archaea almost all fit into one of half a dozen basic shape categories, eukaryotic cells come in hundreds of different shapes. Within our own bodies alone we find neurons or nerve cells, which communicate with each other via their enormously long, enormously thin fibre; red blood cells, which are shaped like squashed balls or hole-less doughnuts; sperm cells that thrash their tails to swim; and oversized adipose cells whose internal volume is almost entirely occupied by a blob of stored fat.

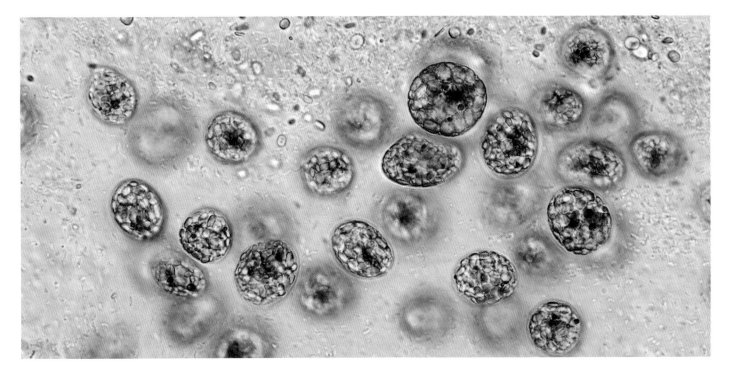

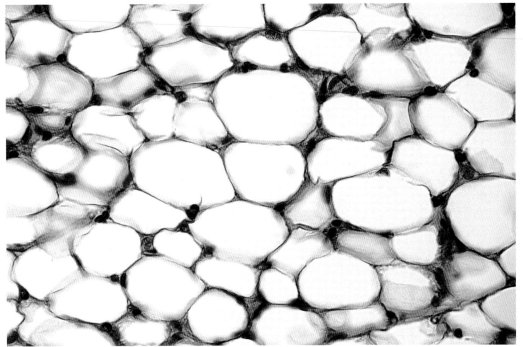

Above *Euglenoid cells are single-celled eukaryotes with photosynthetic capabilities.*

Left *Human adipose cells, used for fat storage. Most of their interior space is occupied by a blob of liquid fat, and they can expand considerably to accommodate more.*

Looking to the plant kingdom, we find very different cell structures, including those with geometric shapes enforced by their rigid walls, which develop in closely stacked formation to provide sturdy support in the form of stems and branches, and root hair cells that extend a long, fine filament into the soil to draw in water and essential minerals. The egg cell or ovum in both plants and animals is an especially large cell, typically much bigger than any other cell in the organism's body, and carries a food supply to sustain the embryo if fertilization occurs.

Even single-celled or unicellular eukaryotes are highly variable in form. No true animals or plants are unicellular, but some unicellular organisms are very animal-like (these are often known as protozoa), and there are also plant-like unicellular organisms (algae), which use photosynthesis. Single-celled fungi do exist and are known as yeasts. Many of us might think of the word "amoeba" when considering single-celled organisms, but this word describes a type of cell shape rather than a particular group of organisms, and amoeboid unicellular organisms may be algae, fungi, or protozoa. Amoebas are characterized by their ability to move by pushing their contents around within the cell membrane, forming projections or pseudopodia ("false feet"), so their shape changes constantly. Some other well-known unicellular eukaryotes include the elongated *Euglena*, with its whip-like flagellum that serve as a swimming tail, the shell-covered foraminifera, and the highly diverse and ecologically important diatoms—photosynthesizers that have rigid and often elaborate and beautiful outer shells that may be triangular, diamond-shaped, wheel-like, or cylindrical.

LITTLE AND LARGE

The species *Caulerpa taxifolia* arguably doesn't belong in this book, despite being a single-celled organism, because it can grow to an extraordinary and very definitely non-microscopic 9.8 ft (3 m) long. The solitary cell of this marine alga has evolved a shape, as well as a size, that recalls a full-blown multicellular land plant, complete with leaf-like projections and a root-like structure to anchor it to the substrate. We explore macroscopic single-celled organisms further on page 120. The smallest known unicellular eukaryote is also a marine alga—*Ostreococcus tauri*. This organism is only 1–3 micrometers long, a similar size to many prokaryotes, and contains one mitochondrion and one chloroplast. It and similar species of algae are known as "picoeukaryotes" because of their tiny size, and like cyanobacteria they are collectively a very important contributor to the Earth's total photosynthesizing production.

Below *A streaked gurnard makes its way over a seabed of the large single-celled alga* Caulerpa prolifera.

// Types of organelles

Eukaryote cells differ most strikingly from prokaryote cells in that they contain a variety of other, smaller structures, each with a distinctive appearance and its own particular function. Between them, the organelles inside a eukaryotic cell handle all the cell's activities, from its self-replication and energy generation to the synthesis of products (such as proteins and hormones) and the storage of nutrients. As we have mentioned, in multicellular organisms many cells are highly specialized and therefore vary considerably in the number and type of organelles they contain.

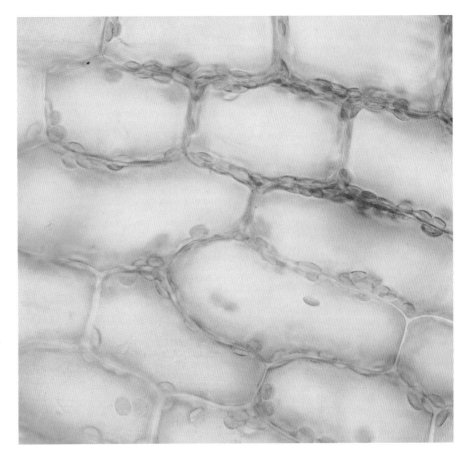

Right *The most visible organelles in plant cells are the green chloroplasts. They also have cell walls, unlike animal cells, which give them a more rigid shape.*

Below *Fully developed human red blood cells or erythrocytes have lost their nuclei and some other organelles.*

Some of the organelles found in most eukaryotic cells are:

Nucleus Usually the most prominent organelle, this structure looks like a large dark spot. It is bound by its own membrane and contains the cell's DNA. It is the location where DNA replication occurs during the process of cell division. Messenger RNA is also built within the nucleus, while a denser region within it, the nucleolus, is where ribosomes are assembled. Some cells have multiple nuclei.

Mitochondria These structures are where the molecule ATP is built, using glucose. The subsequent breakdown of this molecule is what generates energy. As we have seen, mitochondria descend from free-living prokaryotes, and as such they have their own DNA and undergo division independently of the cell in which they exist.

Centrioles This pair of tube-shaped structures are involved with initiating and regulating cell division. When not in use, they are found within a single larger structure called the centrosome.

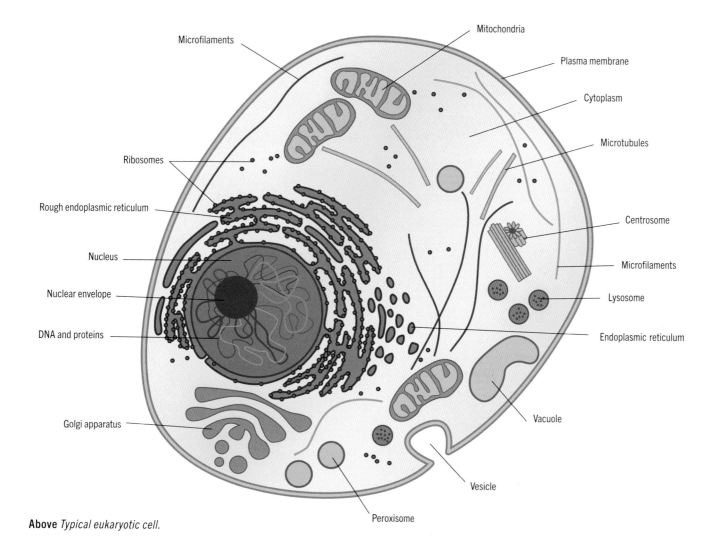

Microfilaments

Mitochondria

Plasma membrane

Cytoplasm

Microtubules

Ribosomes

Rough endoplasmic reticulum

Centrosome

Nucleus

Nuclear envelope

Microfilaments

Lysosome

DNA and proteins

Endoplasmic reticulum

Golgi apparatus

Vacuole

Vesicle

Peroxisome

Above *Typical eukaryotic cell.*

Ribosomes These small organelles exist freely in the cell cytoplasm, and also bound to the endoplasmic reticulum. Their function is protein synthesis. They receive the strands of messenger RNA, built in the nucleus, that carry genetic coding. They "read" this code and assemble the corresponding protein.

Golgi apparatus This is a large membrane-bound structure, formed of a series of flattened disks. Within its compartments the proteins built by the ribosomes are collected and packaged into vesicles for transport out of the cell.

Endoplasmic reticulum Also known as "ER," this large folded, membrane-bound structure, which surrounds the nucleus, is involved with synthesizing nutrients. Rough ER is covered with protein-synthesizing ribosomes, while smooth ER is where lipids (fats) are synthesized from their component fatty acids.

Vesicle This is a quantity of cytoplasm held inside a simple lipid membrane. Vesicles provide storage and transport for various types of molecules, such as proteins.

Vacuole This is a large vesicle. Vacuoles are often highly prominent in plant cells, where they store water and help the cell to stay rigid; they may also store waste products and, in some cases, food supplies.

Flagella and **cilia** Although these structures project from the outside of the cell, they are often considered to be organelles. They provide propulsion to motile cells (for example, the "tail" of a sperm cell is a flagellum).

Other organelles found in some cell types include **plastids** (in plants, where photosynthesis takes place—the best known types of plastids are chloroplasts and all plastids began their evolutionary history as endosymbionts). **Lysosomes** are vesicles that break down large nutrient molecules, the **TIGER domain** is involved in some protein synthesis, and **peroxisomes** carry out the breakdown of the toxic compound hydrogen peroxide, a by-product of some organic reactions.

// Cell division

The process of cell division, also known as mitosis, provides single-celled organisms with a means of reproduction, and multicellular organisms with a way to build body structures and to grow. That an entire large and complex animal grows from a single fertilized egg cell, through repeated rounds of cell division (and within a startlingly short period of time), is one of the everyday miracles of life, and one that we can observe in its early stages under a microscope. In multicellular organisms, many cells become more specialized with successive divisions and reach a "final form" that is fully functional in the particular tissue type or organ where it resides but that is no longer able to divide. However, some other cell types within tissues can continue to divide indefinitely, as can unicellular organisms.

Prior to division, a cell becomes larger, taking in extra nutrients to provide enough material to create two daughter cells. Cell division itself begins with the nucleus moving to the center of the cell, and within the nucleus a copy is made of the cell's DNA. Most eukaryotes have multiple chromosomes, each a strand of DNA, and in normal conditions these long molecules are stretched out and tangled up. Prior to division,

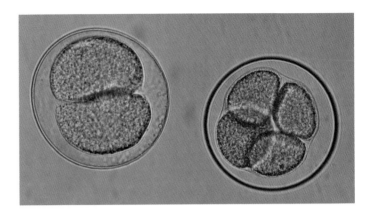

Above *A sea urchin egg cell dividing after fertilization.*

they separate and condense into a more compact form, and then a duplicate set of each is created (this process of DNA replication is described earlier in the book, on page 14). The two duplicates or chromatids are attached to each other in an X-shape—the point of contact is called the centromere. At this time, the cell's centrosome also duplicates itself, with each centrosome containing one of the two centrioles.

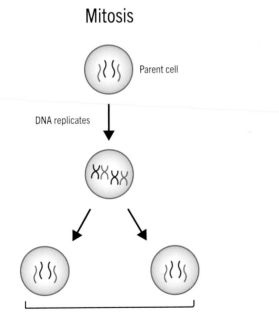

Mitosis

Parent cell

DNA replicates

2 daughter cells

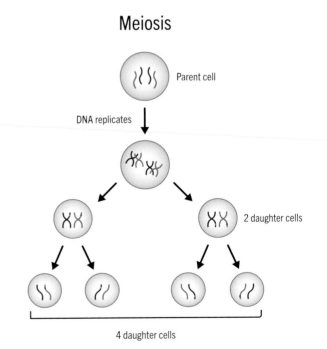

Meiosis

Parent cell

DNA replicates

2 daughter cells

4 daughter cells

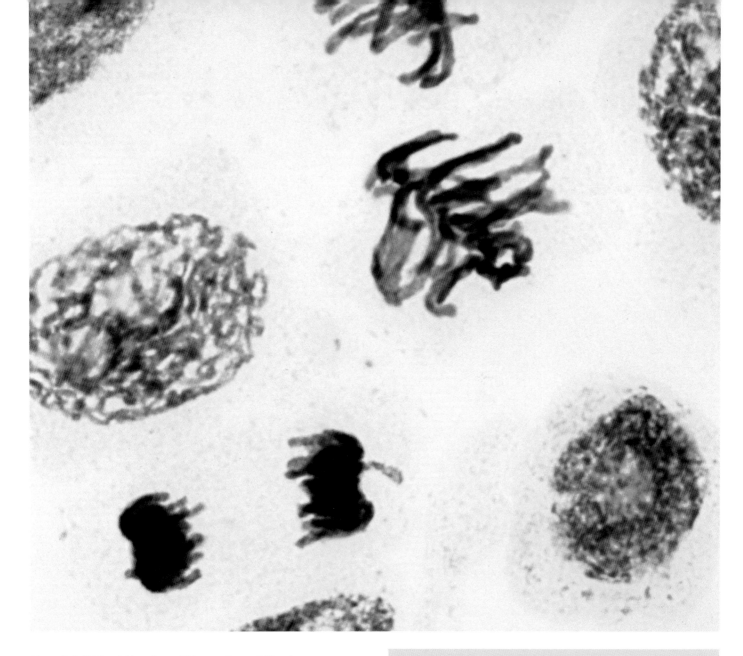

Above *Cell division taking place within an onion root. The chromosomes can be seen condensing and dividing.*

Next, the nuclear membrane breaks down, and the two centrosomes migrate to either side of the now exposed chromosomes. Between them, the two centrioles within the centrosomes form a spindle of microtubules, which attach to the chromosomes. At the center of the spindle the chromosomes are thus lined up and are then pulled apart at the centromeres, with one copy of each chromosome moving away to opposite ends of the cell, creating one full set at each end. Now a new nuclear membrane forms around each set of chromosomes, and as the two new nuclei move away from each other, the cell membrane forms a contractile ring in the cell center, as if "pinching" the one cell into two. Each daughter cell inherits either the original or the new centrosome, along with a share of the other cell organelles.

MEIOSIS

The process described above, mitosis, generates two daughter cells with the same genetic make-up as the parent cell (random copying errors notwithstanding). For sexual reproduction, though, a different process of cell division is required, as the cells it produces need to have only half of the parental DNA. These are called haploid cells (as opposed to a standard diploid cell with a full complement of DNA). Then, two haploid cells from different individual organisms can combine to form a new diploid cell whose DNA is a mixture from both parents. This creates more genetic diversity in a population, and therefore improves the odds that some individuals will have better survival chances. Meiosis is a longer process with an additional round of division, and results in four haploid daughter cells.

// Classification of single-celled eukaryotes

By traditional classifications, eukaryotes fall into four categories—animals, plants, fungi, and ... all of the rest. The first three are typically each given the taxonomic rank of kngdom (one step below domain), so we have the kingdoms Animalia, Plantae and Fungi. The correct classification of those eukaryotes that are not animals, plants, or fungi, though, has long been unsettled. One popular approach is to encompass them all within a fourth kingdom—Protista. However, this "umbrella" grouping does not properly reflect their diversity and evolutionary relationships.

Some systems of classification define the fourth kingdom as Protozoa ("first animals"), which unites a certain group of single-celled eukaryotes that are clearly not plants and not fungi but are somewhat animal-like in both anatomy and behavior. This term is useful for the layperson, but it has also proved to be biologically imprecise. Recently, classification systems with five or more eukaryote kingdoms, and as many as nine, have been proposed. In some of these systems, even Plantae, Fungi, and Animalia are discarded—for example, one concept proposes eight kingdoms, including Opisthokonta, which groups animals and fungi together. It would be fair to say that biologists are yet to reach a consensus over the correct way to classify the eukaryotes at this fundamental level.

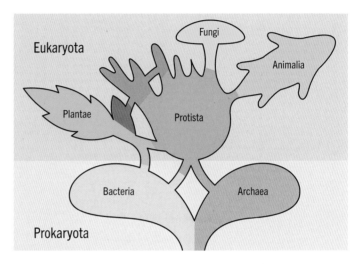

Above *A diagram representing how different eukaryote groups arose from prokaryote ancestors, with Protista shown here as a general grouping of all unicellular eukaryotes.*

Below *This dazzling shoreline bioluminescence is caused by a "bloom" of dinoflagellates, a group of microbes now classified within Chromista.*

Another proposed kingdom is Chromista, which includes single-celled photosynthesizing organisms, but also some multicellular species and some that have lost the ability to photosynthesize. The species classified within Chromista are unlike true plants in that they contain the pigment molecule chlorophyll c, which is not found in other organisms. By this system, some algae are very plant-like and are at least sisters to the true plants, but others are chromists. Some large multicellular seaweed species, such as giant kelp, would be classified in Chromista, along with unicellular groups, such as the diatoms. Most Chromista are plant-like in many respects, but some, such as the dinoflagellates, are much more active than you would expect a plant to be, and even hunt and consume other single-celled organisms. Green algae share the same chlorophyll type as true plants and are sometimes classed within Plantae, or united as a sister group to true plants in a bigger grouping called Viridiplantae. While most green algae are unicellular, some form colonies

with distinct filamentous, leaf-like forms, and others are truly multicellular seaweeds.

The fungi are characterized by the presence of a protein called chitin, found in their cell walls. This protein does not occur in plants, though is found in some animals. Another trait fungi share with animals is their place in the food chain—they are consumers of organic material, rather than being primary producers. So, although we tend to mentally group plants and fungi together because they do not move around of their own volition, it is fungi and animals that are closer cousins (and, just like microscopic algae, some microscopic fungi certainly do move around). Most fungi feed on decaying organic matter but some are parasites and even predators, while others live symbiotically with other organisms (for example, lichens are a union between a fungus and a photosynthesizing organism—see page 57). Unicellular fungi include yeasts, which in some cases can switch to a multicellular form.

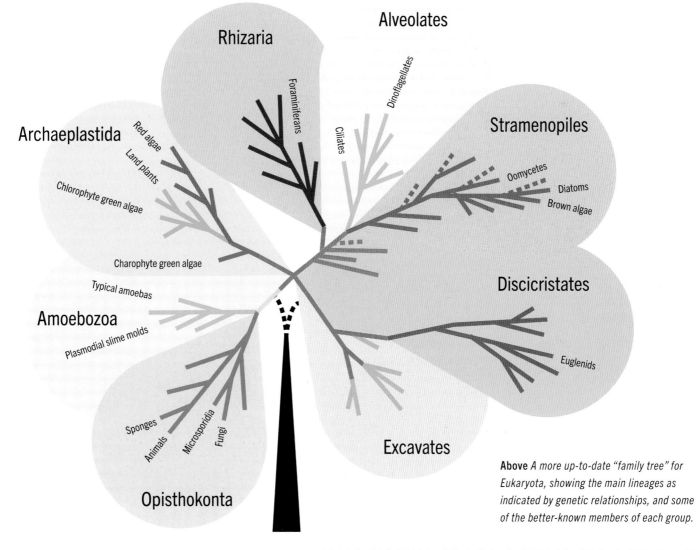

Above *A more up-to-date "family tree" for Eukaryota, showing the main lineages as indicated by genetic relationships, and some of the better-known members of each group.*

// Single-celled algae—types

As we have seen, algae were traditionally considered plants or at least highly plant-like, but today some may be classed in the kingdom Chromista. Other classification systems regard unicellular algae (together with all other unicellular eukaryotes) as members of Protista. In any case, algae have clear plant-like attributes. The green algae are especially similar, having a cell wall and using chlorophyll a or b to carry out their photosynthesis. Like true plants, they also store their food in the form of starch (a long carbohydrate molecule, formed from a chain of glucose molecules).

In some other respects, though, even the green algae are very different to the familiar macroscopic plants we see every day. Virtually all true plants are land-dwelling species that reach their roots down into the soil, and bear leaves on stems that lift them towards the sun. Algae are aquatic, occurring in fresh and salt water (with a few occurring in permanently damp areas on land). Many have no anchorage at all to a solid substrate, let alone any need for roots—unicellular forms can often even swim of their own volition, although many are moved passively instead, by the motion of the water. Even those kinds of algae that are multicellular and are anchored to the seabed do not require strong stems, as they are supported by the water in which they live. Their structure overall is simpler and more uniform than that of true plants.

Below *Although many algae are unicellular, the cells often aggregate to form stringy filaments.*

Left *Not all algae live in water: some species grow over damp substrates on land.*

Below *The delicately pretty red alga* Antithamnion plumula.

The green algae comprise more than 20,000 species. The chlorophytes are single-celled green algae that are mostly aquatic, often forming a film on the water's surface, but can also grow as a green crust on land surfaces such as tree trunks and rocks in moist areas. Some of the chlorophytes form the photosynthesizing component of lichens, while others exist as endosymbionts within certain animals and protozoans, which make use of their photosynthesizing powers. Another important group of single-celled organisms usually regarded as algae are Bacillariophyceae, the diatoms, and Euglenozoa, the euglenids.

RED ALGAE

The red algae, or Rhodophyta, are a group of mainly multicellular plant-like organisms, which carry out photosynthesis but contain a pigment called C-phycocyanin (which gives them their color) as well as chlorophyll. Most red algae are dark red or violet-colored seaweeds with elaborate fronded forms, but one group, Cyanidiophyceae, is unicellular. These cells are very small and simple with minimal organelles (often only a single mitochondrion) and tend to occur in challenging environments such as hot springs with very acidic water.

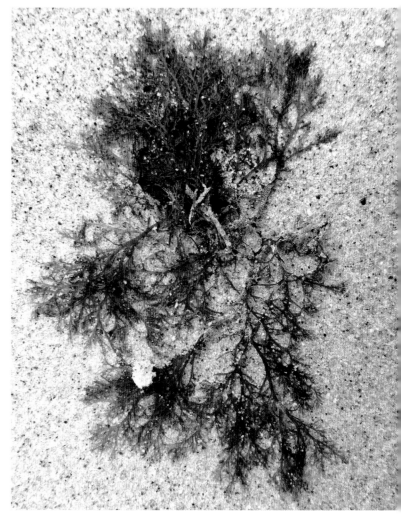

// Single-celled algae—lifestyles

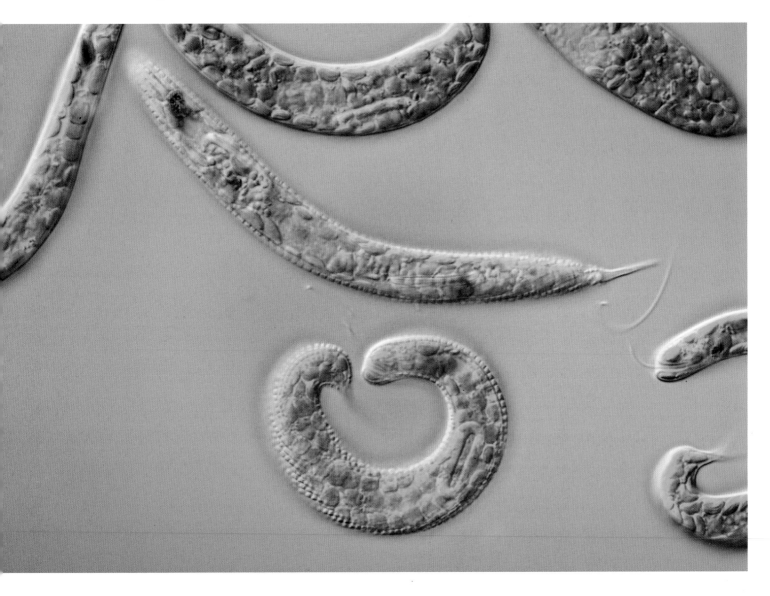

Above Euglena viridis, *a well-known unicellular eukaryote, has a specialised outer layer (pellicle) that gives it great flexibility.*

Although unicellular algae are microscopic individually, many of them are very visible to us because they occur in large aggregations or form quite organized colonies. They make a slimy covering on wet seaside rocks, and floating layers on the edges of ponds, and in general are not that pleasant for us to encounter, but under the microscope they show great beauty of form.

Euglena viridis is a species of unicellular photosynthesizing organism, often considered an alga, and is well known to biologists of all levels, often being studied by newcomers to microscopy. It typically appears elongated and slightly tapered at both ends, though can change its shape to become squatter. It appears bright green because

of its chlorophyll, and like some other algae it has a red-pigmented "eyespot" that filters out some of the light that enters the cell and allows it to orientate itself towards a light source. It has a long hair-like flagellum that it uses to swim. This species is abundant in fresh water and is well studied. Its ability to move freely, and also to sustain itself without sunlight by switching to consuming organic material, led early microbiologists to class it and the other *Euglena* species as a sort of "missing link" between plants and animals. Molecular evidence indicates that the chloroplasts in *Euglena* were originally contained within green algae, and the algae at some point became endosymbiotic with the ancestors of *Euglena*.

SEXUAL REPRODUCTION IN SINGLE-CELLED ORGANISMS

Most unicellular organisms reproduce by cell division. A few produce an endospore, similarly to certain members of Bacteria, and a few more can reproduce sexually—in other words, via two individuals combining their DNA. Sexual reproduction increases genetic variety, which makes a population more resilient (as there are more likely to be some individuals that can cope with any given danger to their survival). With unicellular sexual reproduction, a cell may divide into daughter cells that are themselves gametes of one sex or the other (as with *Volvox*), carrying half a set of DNA from their parent, which they can combine with another gamete's DNA to create a new individual with a full complement of DNA. Another variant, seen for example in yeasts and the protozoan genus *Paramecium*, does not involve gametes at all. Instead, two haploid cells (each carrying a half-set of DNA, the result of cell division through meiosis rather than mitosis) of different and compatible mating strains come together, exchanging genetic material that both participants then use to build daughter cells with new DNA combinations.

Volvox carteri is another well-studied species. This green alga occurs in small, often temporary pools, where it forms spherical colonies that hold as many as 6,000 cells. These cells are mobile and use flagella to move and position themselves. However, when conditions are suitable for reproduction, a few larger, immobile cells develop. These cells, called gonidia, divide multiple times to form a new daughter colony, which is eventually released from within its parent colony and continues to grow. If their habitat is beginning to dry up, sexual reproduction can occur in this species, with colonies producing male and female gametes that fuse to form a durable offspring (a zygote). This form is able to survive in a dormant state until water returns to the habitat.

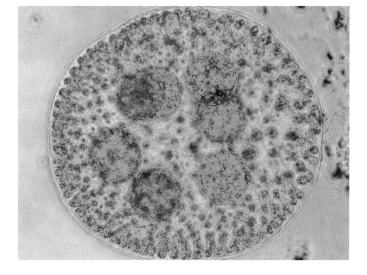

Above *An alga of the genus* Volvox, *in a characteristic spherical colony.*

Left *Cells of algae in the genus* Spirogyra *("water silk") join together to form filaments, and their chloroplasts are arranged in a characteristic spiral pattern.*

// Single-celled algae—photosynthesis

The cyanobacteria were the first organisms to use photosynthesis, and in doing so, they transformed our planet and its atmosphere. All of the other photosynthesizing organisms that have followed in their footsteps have done so not by evolving the same process independently, but by making use of cyanobacteria to do it for them, in one way or another. We consider the Earth's great rainforests to be the lungs of the planet, and indeed their role in our world's ecology is crucial, but most of our oxygen actually comes from the sea. Between them, the marine algae and the cyanobacteria generate as much as 80 per cent of it each year.

After innumerable generations of reproducing inside their symbiotic hosts, the descendants of cyanobacteria are rather different to their free-living cousins. They now function as cell organelles and are known as plastids. Chloroplasts, the familiar green blobs we find inside the cells of green algae and various other unicellular eukaryotes as well as true plants, are one kind, but some other forms of plastids have different functions besides photosynthesis—chromoplasts, for example, make and store pigments, while leucoplasts are used to synthesize other compounds and sometimes for storage.

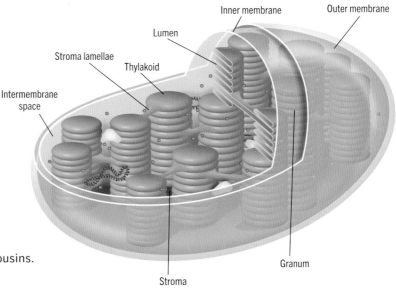

Above *Structure of a chloroplast.*

Below *In ideal growth conditions (lots of sun, and very nutrient-laden water) algae can proliferate to the point where they leave little space for other aquatic life.*

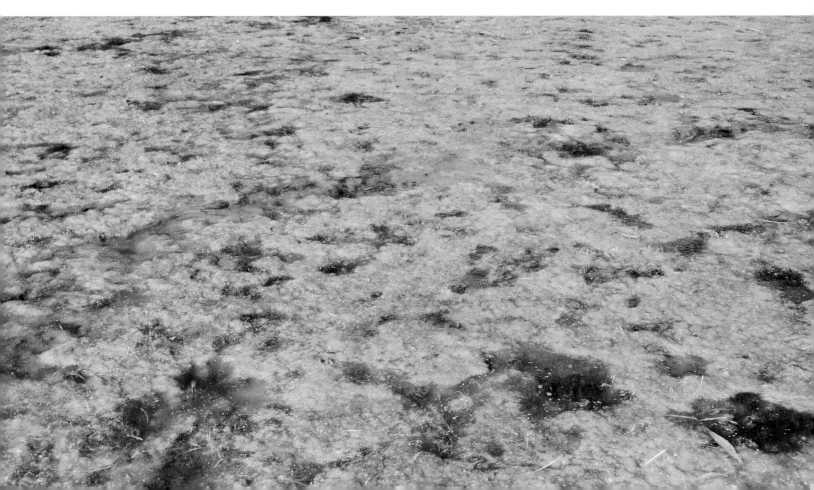

The combination of a cyanobacterium and an archean cell (or perhaps more likely, an early eukaryote that already possessed mitochondria) is a primary endosymbiotic event, and studies of molecular DNA suggest it has occurred successfully only twice in evolutionary history. However, subsequent secondary and tertiary endosymbiosis has since occurred frequently in some organism groups. As we saw on page 86, the ancestor of *Euglena* probably acquired its chloroplasts through combining with a green alga—this is evidenced by the fact that *Euglena*'s chloroplasts have an additional membrane compared to chloroplasts found in green algae (three rather than two).

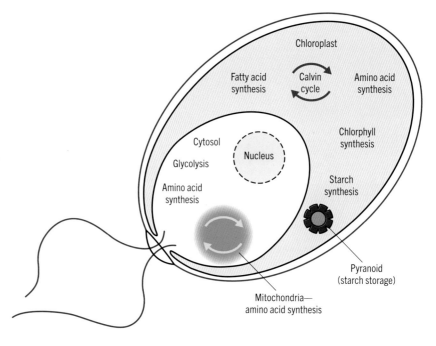

THE CHEMISTRY OF PHOTOSYNTHESIS

As we have seen, the function of photosynthesis is to produce glucose, which plants and algae use as food. The reaction involves carbon dioxide and water reacting together, and releases oxygen as well as generating glucose. Its chemical equation is $6CO_2 + 6H_2O \rightarrow C_6H_{12}O_6 + 6O_2$. But this won't simply happen by mixing together the required reactants. A hefty input of energy is needed for it to work, and this is provided by sunlight. Chloroplasts contain a green pigment called chlorophyll, which absorbs light energy, and this is converted to chemical energy in the form of ATP (see caption on page 142). During this process, water is converted to oxygen and hydrogen. The oxygen is released, while the NADPH captures the hydrogen component. The next stage of the reaction, which does not require sunlight, involves the chloroplast fixing carbon dioxide (powered by energy supplied by ATP) and using it plus the hydrogen carried on the NADPH to build glucose molecules. These glucose molecules can then be built into chains to provide food storage (starch) or modified to form amino acids (the basis of proteins) or fatty acids (the basis of fats, or lipids).

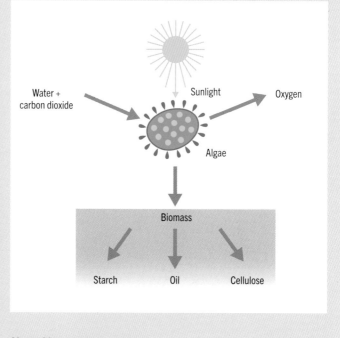

Above *Photosynthesis in algae.*

// Single-celled algae—movement

When you watch *Euglena* in action under a light microscope, you'll be struck by how speedily they can dart across your field of view. Other algae such as dinoflagellates are also able to get around at an impressive pace. These and most other algae swim primarily with the use of a flagellum or pair of flagella, though some have more than this. We have also seen that some prokaryotes have flagella and use them as "swimming tails" although they are structurally and functionally different to the flagella of eukaryotes.

Organisms that use flagella to move are collectively known as flagellates. Flagellate algae may use one flagellum (as in *Euglena*, although it does have a much shorter, non-swimming second flagellum) or two of equal length. When there are a pair, they are usually side by side, as in the spherical green alga *Chlamydomonas*, and are located at the end of the cell that is at the front when the cell is in motion. The cell's membrane extends over the outside of the flagellum, and its interior contains microtubules that, in cross-section, form a circular arrangement. These tubules can slide over each other, but the extent of that movement is limited by protein connections between the tubules. Consequently, the flagellum can bend and flex. During swimming, it moves in a whirling, propellor-like motion.

In *Chlamydomonas*, with its two equally long flagella, the cells spin as they swim. This movement allows it to assess light levels all around itself, as the position of its light spot shifts around with each swimming action. It can then modify its stroke pattern, by using more energy to drive the motion of one flagellum and less for the other, to steer itself towards the light. Close observation has revealed that these and other

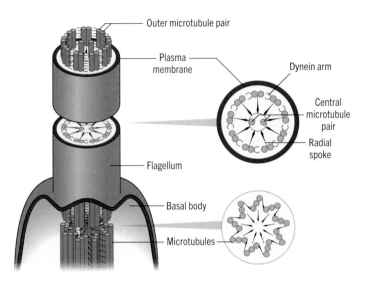

Above Structure of eukaryote flagellum.

algae can modify their "stroke" considerably. The species *Pyramimonas parkeae* moves its two pairs of flagella in an alternating motion, likened to the trotting of a horse, while its structurally similar cousin *Pyramimonas tetrarhynchos* performs a three-beat "gallop" and sometimes also a "pronk," when all four flagella push the same way at the same time to make a quick directional change. Some algae have eight and even 16 flagella so could potentially move in an even greater variety of ways.

MOVEMENT BY PELLICLE

Euglena and its relatives do not have a cell wall like most algae but instead possess a specialized outer coating called a pellicle. This protein-based layer contains sets of microtubules, arranged in strips that can slide over each other, giving the cell a faintly striped appearance under high magnification. As with the flagellum, the sliding microtubule structure provides flexibility, allowing the cell to quickly transform its shape from a slim, elongated form to an almost spherical shape, and to pulse back and forth between the two shapes. This works with the flagellar movement to add to its swimming ability, although in some of the euglenids the flexibility has been lost.

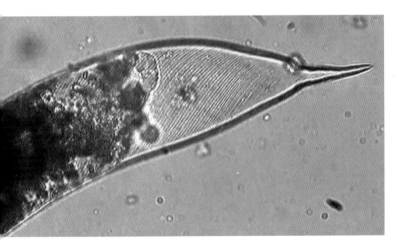

Above In extreme close-up, the striations on a euglenid's pellicle are visible.

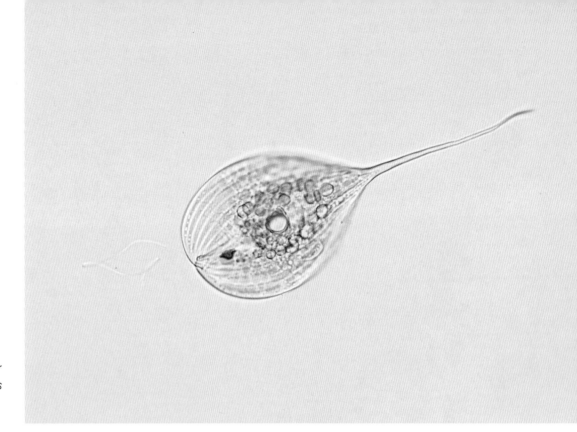

Right Phacus, *a genus of leaf-shaped euglenoids, usually have one full-sized flagellum, but in poor conditions the flagellum disappears and the cells grow larger to allow for more nutrient storage.*

Left *Many algae bear two flagellae, which sit side by side.*

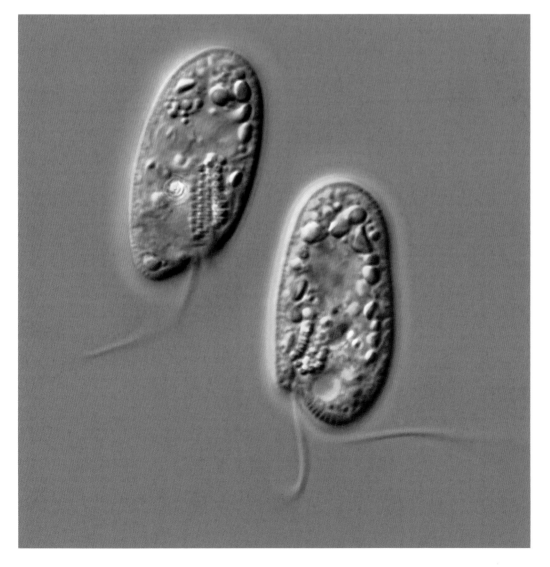

// Diatoms

The lush rainforests of South America, full of sky-scraping trees, owe much of their vigor to microscopic sea organisms that died long before, thousands of miles away. Saharan dust, carried on prevailing winds, falls with the rains into the Atlantic and from there is carried up the river systems of the Atlantic seaboard. This mineral-rich dust is full of the remains of diatoms, microscopic algae that lived in abundance in North Africa thousands of years ago, before the Sahara had formed and when the landscape was dominated by vast freshwater lakes.

Diatoms are tiny algae whose intricately textured, hard silicate cell walls or frustules serve as protective shells and give them a glistening, glassy appearance. These microscopic photosynthesizers are an important constituent of the ocean's phytoplankton ("plant-like plankton," as opposed to the non-photosynthesizing, animal or animal-like zooplankton) and contribute a large proportion of the planet's atmospheric oxygen each year, particularly in spring and autumn, when populations increase dramatically (bloom). They also remove huge amounts of silicon from sea water, as this is a key constituent of their shells.

They move only passively, carried on ocean currents (although some do stick to a substrate). The turbulence at the sea surface helps to keep them close to the sunlight, and in some cases their complex shell shapes help to keep them afloat too. Some species that attach to a substrate are capable of active movement, as they produce a mucus that allows them to move across a surface. This mucus also allows them to attach to other diatoms to form a colony. Some species are capable of sexual reproduction under certain circumstances, and the male gametes they produce have flagella and actively swim in search of female gametes.

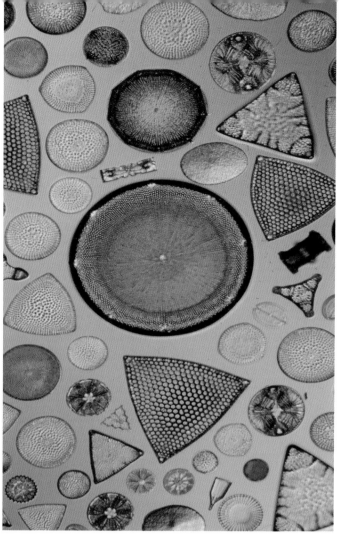

Above *Diatoms have a glassy appearance and come in a variety of shapes.*

Left *Red Saharan dust, formed by long-dead diatoms, is swept westwards by prevailing winds and can cause challenging air conditions.*

SHRINKAGE

When diatoms reproduce (in the usual asexual method of cell division), the cell size of the daughter cells is smaller on average than the parental generation, because their new shells grow inside the rigid parental shell. This trend continues over successive generations, with the cells becoming progressively smaller and smaller. Given that normal-sized diatoms are already classed as microalgae, this shrinkage cannot continue for very long, and eventually the dividing cells produce gametes, which come together to form specialized cells called auxospores. These cells have much less silica in their walls so can grow larger over time, rather than dividing, although they are also capable of remaining dormant for extended periods. When conditions favor their growth, they eventually reach the correct cell size in the population, and then begin to divide to start the cycle again.

// Chromista

Some of the algae we have explored in the previous pages, including the dinoflagellates and diatoms, are often now classed in a kingdom of their own—Chromista. This kingdom also includes some organisms that do not photosynthesize and are not considered to be algae; for example, the organism *Plasmodium*, which is the causative agent of malaria, and the ciliates. Some of these we discuss later in the book under their older, traditional grouping of Protozoa.

Two traits define an organism as a chromist, though it only needs one of them to qualify. The first is that their plastids contain chlorophyll c (though they may also have chlorophyll a) and have four membranes around them (rather than two, as in most photosynthesizers, or three as in the euglenids). The second is the presence of cilia with a distinctive structure. The plastids found in chromists are believed to have been acquired via a secondary endosymbiotic event, involving red algae (rather than green algae as in the euglenids, although secondary symbiosis involving green algae is also likely to have occurred in some chromists).

Besides those already discussed, other unicellular, microscopic chromists include the golden algae, which appear vivid yellowish under a microscope. Some of these algae are motile, swimming with one long flagellum, while others settle down and form colonies that take on an elegant branched form, like a miniature tree. One species, *Prymnesium parvum*, is notorious as a cause of mass fish die-offs when it blooms, as it releases a deadly toxin called prymnesin. These "fish kills" have been documented from both sides of the Atlantic. Another well-known group classed in Chromista is *Paramecium*, a genus of organisms that are common in all watery environments, and often find themselves under the microscope in school biology labs. *Paramecium* species are ciliates, swimming via beating motions of numerous short cilia, and appear slipper-shaped because of an indentation called the oral groove, into which food particles are swept by the action of the cilia.

Other chromists are multicellular and include the brown algae. Most of the brown seaweeds we encounter in temperate and polar seas and shores, such as various kelp and wracks, belong to this group. The brown algae are close cousins to the yellow-green algae, which are mostly unicellular and occur primarily in freshwater habitats.

Below Rhabdonema, *a golden alga classed within Chromista.*

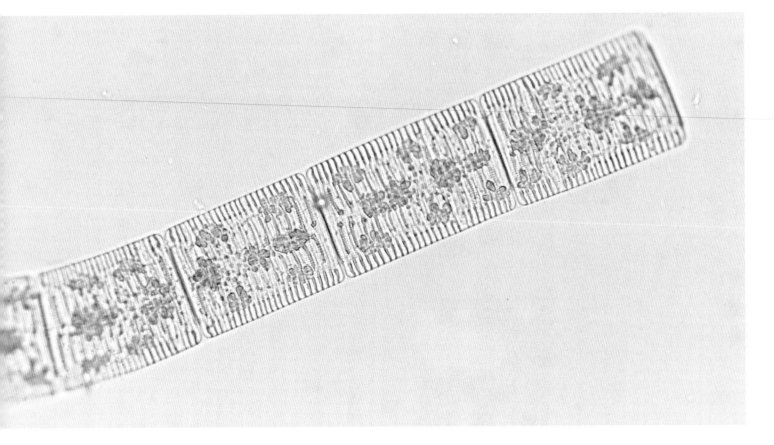

MONOPHYLY

The goal of any system of biological classification is to accurately capture the evolutionary relationships between different kinds of living things. If the system is correct, then all defined taxonomic groups or clades are monophyletic—they include an ancestral form, and all forms (living or extinct) that subsequently evolved from that one common ancestor. You might look at it as one branch plus all of the twigs that grew from that branch. An example of a monophyletic group would be Hominidae—the great apes. This grouping includes chimps, gorillas, orang-utans, and humans (with multiple species of all). The genetics of the living species in the group confirm monophyly. However, establishing monophyly is not always straightforward, even when comparing genomes. The extensive ongoing disagreement over how to classify many kinds of organisms highlights this difficulty! The kingdom Chromista was proposed as a monophyletic grouping but some biologists now consider it to be polyphyletic—a group with multiple common ancestors rather than one. If so, then it cannot be regarded as a biological kingdom.

Above *A fish kill at the River Odra, Poland, caused by a bloom of the alga Prymnesium parvum.*

Below *Examples of a polyphyletic and a monophyletic grouping within the primates.*

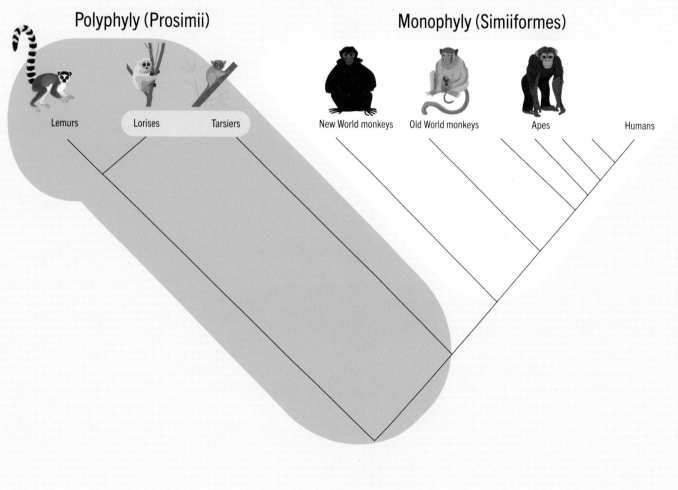

Polyphyly (Prosimii)

Lemurs   Lorises   Tarsiers

Monophyly (Simiiformes)

New World monkeys   Old World monkeys   Apes   Humans

// Single-celled fungi (yeasts)—types

Unlike Chromista, the fungi are quite widely accepted as a correct monophyletic grouping. DNA studies suggest that their earliest common ancestor probably lived at least 1.2 billion years ago. It was most likely an aquatic unicellular organism that swam with the aid of flagella, so was not so different to tiny algae and protozoa, at least at first glance. The earliest fossils of fungi that resemble modern multicellular land-dwelling species date back some 635 million years, but fungi, unlike many other organisms, have no hard material in their structure so fossil evidence is thin on the ground. The presence of the protein chitin in their cell walls sets them apart from plants and animals alike (plant cell walls have no chitin, and animal cells lack cell walls).

Above *This peculiar fungus is a morel, a multicellular close cousin to the single-celled ascomycete yeasts.*

Below *How yeast cells reproduce by budding.*

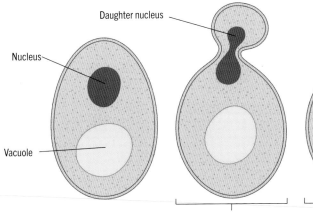
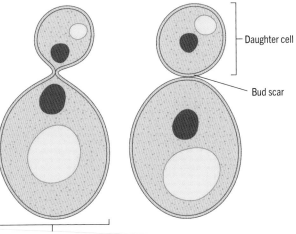

Daughter nucleus

Nucleus

Vacuole

Daughter cell

Bud scar

Bud formation

Nucleus migration

Today, about 1 per cent of all known fungal species are unicellular, and collectively these 1,500 or so species are known as yeasts. They are not, though, all members of a single lineage, but examples can be found in a range of quite distinct groups of fungi. It is also clear, therefore, that modern yeasts are not straight-line descendants of the first single-celled fungi. Instead, they are descended from multicellular ancestors, and some types can readily switch to a form of multicellularity in certain conditions, growing the branching threads (hyphae) that form the bulk of the "bodies" of multicellular fungi. Most yeast cells are very small, no more than 4 micrometers in diameter.

Yeasts obtain their energy from consuming organic material (especially sugars) rather than by absorbing sunlight, and all are capable of aerobic respiration, although they may also use anaerobic respiration and generate a range of by-products in the process (some of them commercially useful). They are abundant in all kinds of habitats, on land and in water, even the deep sea. Some live in symbiosis with other organisms, while others are parasites and some can switch from living commensally on an organism to becoming pathogenic. The usual asexual reproduction method is budding, in which an outgrowth develops on the cell, and when the parent nucleus divides the daughter nucleus migrates into the bud. The bud eventually reaches full size and splits off but usually remains stuck to the parent cell. This is similar to the way that the hyphae grow in multicellular fungi. Many yeasts can also reproduce sexually, which multicellular fungi also do, with haploid cells of two different "mating types" fusing together to form a spore.

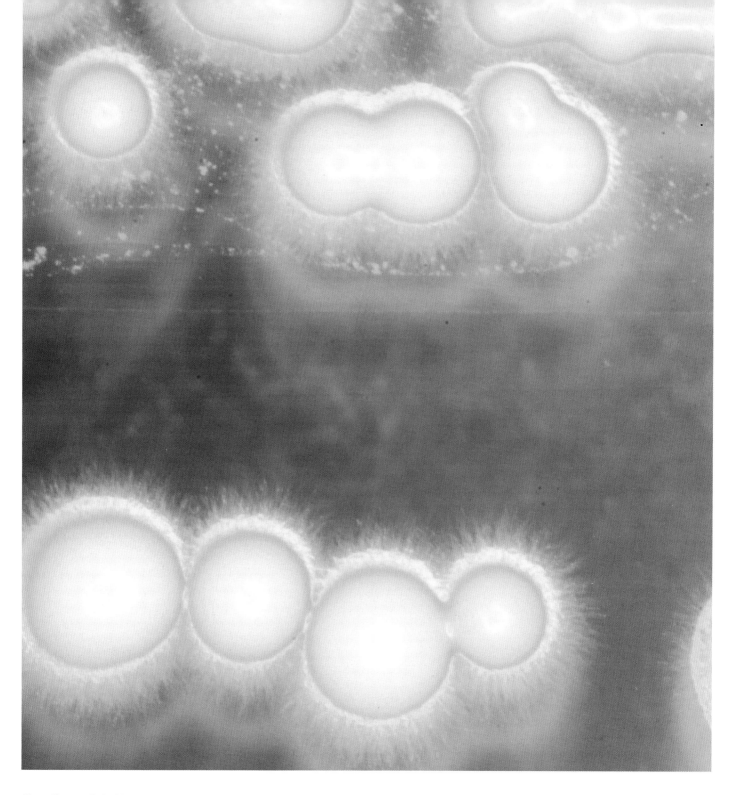

Above *Yeast cells budding under the microscope.*

FAMILY RELATIONSHIPS

Most known yeast species belong to the taxonomic order Saccharomycetales—the ascomycete yeasts. This group includes the species that are most familiar to us—some cause disease in humans and other species, and also in some economically important plant crops, but other species in the group are used in industrial and medical processes and have great commercial value. Other yeasts, though, are represented in the group Basidiomycota, the "higher fungi." The familiar mushrooms and toadstools that we find in the forests and the supermarket are members of this group, as are various intriguing microscopic, parasitic species.

// Yeasts—lifestyles

Considering that a yeast cell has a rigid wall and is not able to move of its own volition, it's surprising and impressive that these organisms have diversified into so many different ecological niches. One thing that yeasts can do quickly, though, is adapt to changing conditions. A colony can grow at high speed when conditions are right and may produce durable dormant spores or switch from asexual to sexual reproduction when nutrients are scarce, producing cells of two compatible "mating types," each with a half-set of DNA.

Yeasts live in the soil and on fruit-bearing plants, proliferating on the fruits' skins and consuming the flesh as they mature and decay. This is not to the plant's detriment, as fruit needs to be consumed in order for the seeds within to be released. Their sugar-loving ways have also led them to colonize the nectaries of flowers, and the honey-stomachs of bees. These relationships could be mutualistic, benefiting the host as well as the yeast—at least one case, the presence of active yeast in flower nectaries, has a warming effect, which

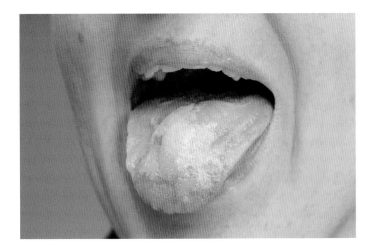

Above *Infection of* Candida albicans *or thrush on a human tongue.*

Below *Yeast activity raises the temperature in the flowers of stinking hellebore, allowing their bee-luring scent to travel further.*

allows the flower's scent to travel further and potentially attract more pollinating insects.

Some yeasts are actively predatory, attacking and consuming other yeast species, while others use toxins to kill off competing yeast colonies. One of the species that we know best, *Candida albicans*, can occur naturally and harmlessly on human skin but can build up and become pathogenic, causing the uncomfortable condition known as thrush. If it enters the bloodstream, it can also cause a more dangerous disease, invasive candidiasis, which can be lethal. Immunocompromised patients are especially vulnerable to it, and patients on antibiotics are also susceptible, as the drugs kill off bacteria and thus "make space" for yeasts to thrive more successfully. Another yeast, *Malassezia*, is also part of the natural community of microorganisms on human skin, but if the skin becomes inflamed then the presence of *Malassezia* can significantly worsen the problem.

RAPID MULTICELLULARITY

Multicellular life has evolved independently many times in the history of life on Earth, in contrast to, say, primary endosymbiosis, which is thought to have occurred very few times. When a colony of single-celled organisms begin to differentiate, with different cells performing different functions, this is the first step towards true multicellularity. By applying a selective pressure that favored colony formation, scientists have been able to induce the beginnings of multicellularity in a yeast species in as little as 60 days. These cells had begun life as a population of genetic clones, produced through asexual budding, but they quickly formed snowflake-shaped colonies that, like true multicellular life forms, had distinct life stages, and reproduced miniature new snowflake colonies rather than random new individual cells.

Below *Cells and hyphae of* Malassezia furfur, *a fungus that causes various skin problems in humans.*

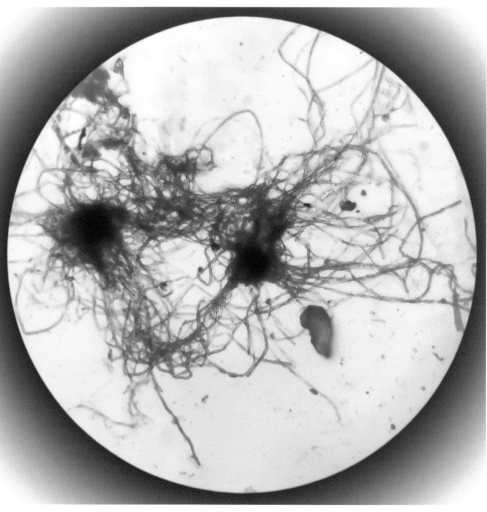

// Yeasts—uses

Next time you enjoy a slice of buttered toast, or a foaming pint of beer, remember to thank the unicellular fungi that helped to create that experience for you. Yeasts are key to producing these products, because of the way that they metabolize their food. Flatbreads are made without yeast and do not puff up and develop a spongy interior the way that yeasted breads do, which gives a clue as to the nature of their action during bread-making, while beer made without adding yeast would be a somewhat beer-flavored but alcohol-free beverage.

Both baker's yeasts and brewer's yeasts are usually the species *Saccharomyces cerevisiae*, but two distinct strains. The action of the yeast is the same, though—it consumes sugars present in the grain component of the bread dough or the mash of barley and hops that is used to make beer. The reaction releases carbon dioxide and ethanol alcohol as by-products. In bread dough, the bubbles of carbon dioxide cause the dough to puff up during the proving process (when the dough is left in a warm place for a couple of hours to encourage the yeast's activity). Then, during baking, the spongy structure is "set" as the gluten protein in the dough goes from stretchy to stiff, while the heat of the oven causes virtually all of the ethanol to evaporate. Both bubbles and ethanol are desired and retained in beer, and the yeast is given much longer (usually at least two weeks) to work its fermenting magic. Yeast used during wine-making too.

Right *The foamy head of a pint of beer is the result of brewer's yeast releasing carbon dioxide as it ferments sugars.*

Below *Brewing beer starts by heating a "mash" of grains, which releases their sugars for the brewer's yeast to digest later.*

We also use yeasts to produce ethanol on a large scale, to produce ethanol fuel, and in some non-alcoholic beverages to provide carbonation (the process is halted before any significant amount of ethanol has been generated). Rich in B vitamins, yeast is used as a food in its own right, as flaked "nutritional yeast" or in savory spreads like Marmite. Added to aquariums, it can provide underwater plants with an extra supply of carbon dioxide.

The potential medicinal uses of yeast have long been of interest to us. For example, brewer's yeast is a purported alternative treatment for conditions as diverse as eczema, gout, and diabetes (the latter because of its ability to metabolize sugars). In the field of biotechnology, yeast is

Above *Bread being proved. Yeast is active in this part of the baking process, fermenting sugars and releasing the carbon dioxide that makes the dough swell.*

showing great promise as a way of synthesizing certain drugs that are currently obtained from plants and other organisms that are less easy to manipulate in the lab. Genetically engineered yeasts have already proved able to make synthetic opiates (usually made from poppies), and D-lysergic acid (DLA—a medicine used to treat dementia, which is currently made from a cereal-infecting fungus called ergot). Production is not yet possible on anything like a large enough scale to meet our needs, but progress is highly promising.

Below *Anaerobic respiration in yeast.*

$$C_6H_{12}O_6 \longrightarrow 2C_2H_5OH + 2CO_2$$

// Oomycetes

When is a fungus not a fungus? When it turns out to be an oomycete, or "water mold" as these enigmatic microbes are sometimes known. This group of non-photosynthesizing organisms includes unicellular and multicellular forms, and their similarity to fungi is apparent in that they produce fine, hair-like growth filaments, resembling the hyphae of true (multicellular) fungi, and some produce chitin. Many are also saprophytic (feeding on decaying organic material). However, once their DNA was unravelled, they proved instead to be closer cousins to brown algae, diatoms, and other members of the group Chromista. It is likely, therefore, that they descend from a photosynthesizing ancestor, but have switched to becoming consumers rather than producers, and that their fungus-like traits developed because of convergent evolution.

Oomycetes reproduce by generating spores, which are mobile and will move through water in response to sensing the presence of certain chemicals, indicating a nearby food source. A few others make spores that are distributed by the wind. They can reproduce sexually or asexually, with sexual reproduction occurring at times when resources are becoming scarce. This occurs through hyphae bearing male gametes coming into contact with others bearing female gametes, and results in a different kind of spore (oospore) that can lie dormant to survive unpromising conditions.

While many oomycetes live in water or soil and are important components of nutrient-cycling through their breakdown of dead organisms and other organic material, others are parasites or pathogens of quite alarming virulence, particularly of plants and fungi. The group includes *Phytophthora infestans*, the cause of potato blight, which caused a devastating famine in Ireland in the mid-19th century. This parasite can also attack tomatoes and other related plant species. Another, *Phytophthora ramorum*, causes the disease sudden oak death, which has had a dramatic impact on woodlands in North America and Europe since the late 1990s. A few forms produce disease in animals, such as *Pythium insidiosum*, which causes the rare but lethal flesh-decaying disease pythiosis.

Below *Potato blight, a disease responsible for a great deal of human misery, is caused by oomycete infection.*

Right *Oomycete diseases such as late blight in tomatoes can devastate a harvest.*

Below *The Irish potato famine of the 1840s–50s killed about a million people and displaced even more, having a dramatic affect on the country's population.*

RECRUITING A KILLER

Although oomycetes are associated with some very serious plant diseases, the species *Pythium oligandrum* has a rather different role, as a plant protector. It is a parasite of at least 20 species of plant-attacking fungi and other oomycetes as well, including members of the genus *Phytophthora* that are so notorious as plant pathogens. The presence of *P. oligandrum* in plant roots helps to keep plant pathogens in check, and also stimulates the plant to grow with more vigor and to improve its own defense responses against pathogens. This oomycete was licensed for use as a biological pest control agent in 2007 and is supplied in oospore form in a carrier liquid, to be added to soil.

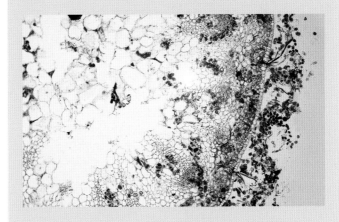

Above *The oomycete* Peronospera *on its host's cells. Some members of this genus are used in biological pest control.*

// Microsporidia

The microsporidia are single-celled organisms that were traditionally classified as primitive forms of protozoa. However, conversely to the oomycetes, their genes reveal them to actually be close cousins to the fungi, perhaps even rightly belonging in the Fungi kingdom, despite lacking certain typical fungal traits, although they do reproduce by forming spores. All known forms are parasites or symbionts of animals, usually residing in the intestinal tract, and most have only a single host species, while some rely on more than one host at different stages of their life cycle.

The cellular structure of microsporidia is distinctive and rather simple. Their genomes are short and simple, carrying fewer protein-coding genes than any other known eukaryotes. They do not have mitochondria, but instead have mitosomes, which are smaller organelles that are believed to descend from mitochondria but no longer have any DNA of their own (and carry out fewer functions than full-blown mitochondria do). They also lack any means of movement. Given their evolutionary path, it is apparent that these are features that they have lost, along with photosynthesis, as they adapted to a parasitic way of life with a host species fulfilling many of their needs. Their reproduction method, though, has required considerable adaptation along a relatively complex pathway in some cases, with both asexual and sexual stages as well as multiple hosts. The spores they form are exceptionally resilient and may spend a long time in a dormant state before being taken up by the correct host species.

At least 14 species of microsporidia are known to be able to infect humans, typically via drinking contaminated water. The disease they cause, microsporidiosis, is not noticeably different between species, and is not usually severe, although it can take hold more seriously in people with compromised immunity, causing diarrhoea and weight loss. They are especially prevalent in insects and can pass from a female insect's gut into her ovaries, thus passing on to the eggs she lays.

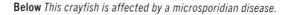

Below *This crayfish is affected by a microsporidian disease.*

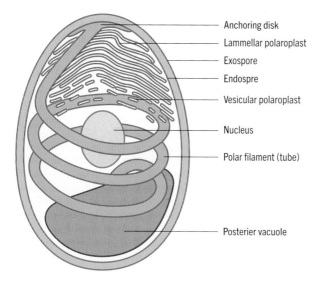

- Anchoring disk
- Lammellar polaroplast
- Exospore
- Endospre
- Vesicular polaroplast
- Nucleus
- Polar filament (tube)
- Posterier vacuole

HYPERPARASITISM

Being parasitized is a fact of life for most organisms. Parasites are distinguished from commensal organisms because they actively harm their host, stealing resources and often causing cell damage, even if this is not very serious. Some 300 different organisms are host-specific parasites of humans, for example, including some microsporidia alongside much larger and more complex organisms, such as tapeworms and scabies mites. None of them is an appealing prospect, but parasites don't have it all their own way, because many of them host parasites of their own. Parasites of parasites are known as hyperparasites, and several microsporidia pursue this lifestyle.

Above *Diagram of a spore of a microsporidian*

Below *Domestic honey bees often suffer from the disease nosema, caused by one of two species of microsporidia.*

// Animal-like protists (protozoa)—types

The first scientists to peer through a lens into a sample of pond water and observe the tiny moving creatures within believed they were looking at miniaturized animals. This is unsurprising, because we have long observed that animals move of their own volition, while other living things do not. The name Protozoa or "first animals" was coined in the early 19th century to describe these single-celled organisms. We recognize today that many moving unicellular eukaryotes belong within other lineages, rather than the one that gave rise to the animals. The term remains widely in use, though, usually as an informal way to describe those single-celled eukaryotes that do show the most animal-like traits, such as being motile, non-photosynthesizing, and lacking the cell wall possessed by plants and fungi (instead, they just have a cell membrane, so compromise durability but gain flexibility). The cell membrane allows water and some other molecules to pass in and out through osmosis, so the organism needs a sensitive osmoregulatory system so it does not take on or lose too much water.

The family Amoebidae share their name with the "amoeboid" cell type, whereby the cell membrane can distort dramatically to form projections known as pseudopodia. These allow the organism to move over a substrate, in some cases in a speedy motion that almost resembles the scuttling of a many-legged animal, and also to engulf its food. The true amoebas are generally free-living, water-dwelling and predatory, capturing bacteria, algae and even some very small multicellular eukaryotes. Some members of the group are very large for unicellular organisms and visible to the naked eye, their cells containing a host of organelles including multiple nuclei. Some groups (the testate amoebae) secrete or build (from sediment particles) a sturdy shell for protection, from which they extend their pseudopodia to feed.

The ciliates, a large group now sometimes classified within Chromista, include the well-known *Paramecium*. Their swimming motion is driven by the beating of many tiny cilia on the cell membrane. Their other traits include a double nucleus (with one used to regulate cell activities in general while the second, smaller "micronucleus" houses the genetic material). They are also predatory in the main but unlike amoebas, which can engulf prey at any point around the cell membrane, most ciliates have a defined "mouth."

Cercozoa is a large group of protozoa that either get around by pseudopodia or have flagella for swimming—a few use both. Most live in water but some are soil-dwellers and others live as parasites, in plants, algae, oomycetes, or marine invertebrates. Some of those that are plant parasites are economically significant because they target widely grown crop plants. Another group of protozoa, the diplomonads, swim with multiple flagella and parasitize animals, including humans. They are also among the protozoa that regularly reproduce sexually. The parabasalids are also flagellates, and (rather like the microsporidia) have a simple structure relative to their ancestors, including a lack of mitochondria. These organisms are mainly symbionts of animals.

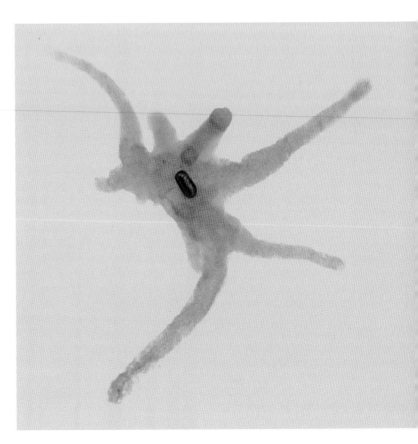

Right Entamoeba histolytica, *the causative agent of the unpleasant illness known as amoebic dystentry.*

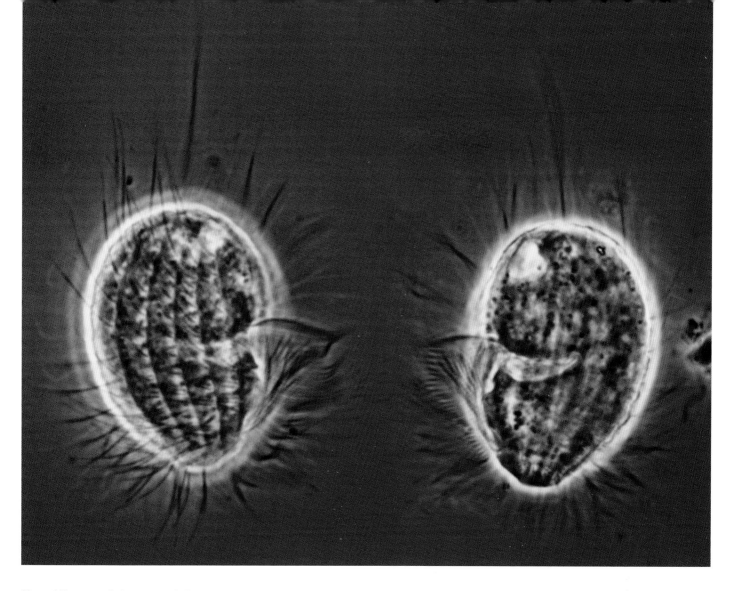

Above *Ciliates use their numerous hairs or cilia to propel themselves.*

Below *Anatomy of a* Paramecium. *Note the rudimentary mouth and gullet.*

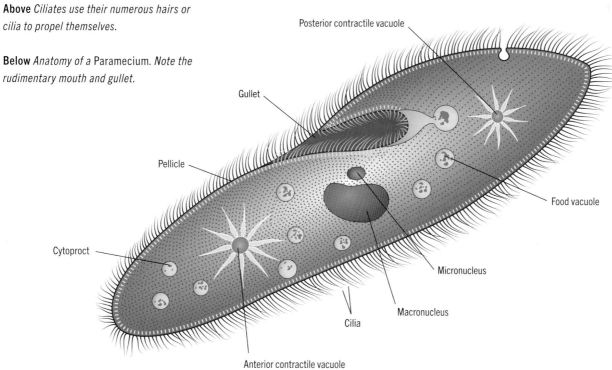

Posterior contractile vacuole

Gullet

Pellicle

Cytoproct

Food vacuole

Micronucleus

Macronucleus

Cilia

Anterior contractile vacuole

// Protozoa—lifestyles

These organisms, as we have seen, include predators and parasites, and others that live within a host but provide benefits. They are eaters of organic material but are far from passive in their efforts to obtain sustenance. Those that are predators are as active and effective in their hunting role as any bird of prey or big cat, and their "kills" are often barely smaller than themselves.

The means by which amoebas typically capture their prey is called "phagocytosis." We can see similar cell behavior going on in our own blood, with amoeboid white blood cells swallowing up pathogenic bacteria and the like, and containing them in a pocket within their cell membranes. The membrane seals completely around the prey, bringing it into the cell in a membrane-bound bubble called a phagosome, while the outer membrane re-forms as a continuous surface. Enzymes, built elsewhere in the cell, pass through the membrane of the phagosome to digest the captured prey. This would not be an option for any organism with cell walls surrounding the cell membrane—the membrane on its own, though, has the necessary flexibility and permeability. Some other protozoa do not engulf their prey but instead fuse to it and then allow digestive enzymes to flow through both prey and host membranes.

Many parasitic protozoans can be free-living, although only proliferate once they have made their way into a host. They may move between hosts via water, soil, or another damp substrate (for example, the organism *Trichomonas gallinae*, which causes disease in birds, is often spread via garden bird feeders that have been contaminated by the saliva or droppings of infected birds). The parasitic protozoa genus *Toxoplasma* has been noted to cause behavioral changes in its hosts that increase the chances of it being passed on to new hosts—for example, infected mice lose their fear of the scent of cats and so are more likely to be caught and eaten, allowing the parasite to find a new home in the digestive tract of the cat.

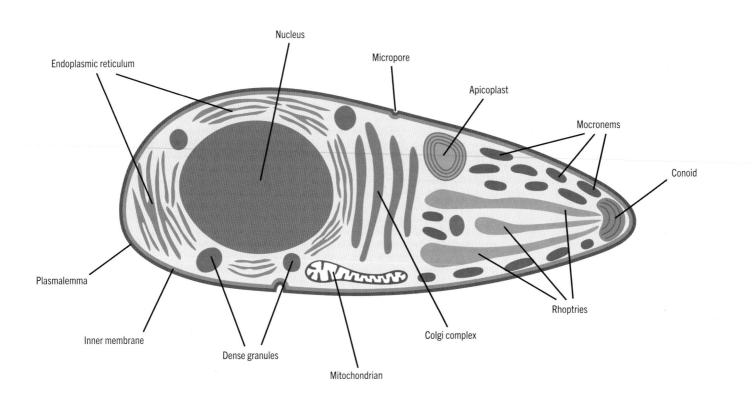

Above *Anatomy of the parasitic* Toxoplasma.

TERMITE'S FRIEND

Some representatives of two flagellate groups, the parabasalids and the oxymonads, occur in the digestive tracts of termites and nowhere else. They carry out digestive breakdown of the tough cellulose in the wood that is eaten by their hosts. Ruminant mammals like cows also have populations of protozoa in their rumens, alongside bacteria, to aid digestion. As we have seen, plants make their own glucose through photosynthesis, and they make starch from this. A starch molecule is a polymer of glucose—a series of glucose molecules joined together—and this is how a plant stores its food. If the glucose molecules are arranged differently, though, they form a different polymer—cellulose. This is used by the plant as a structural component and it is very sturdy, to help protect the plant from being eaten. Animals that are able to digest it can often only do so with a lot of help from specialized microorganisms.

Right *Termites can do extensive damage to wood-framed buildings, assisted by the protozoa in their guts.*

Below *Diagram showing phagocytosis.*

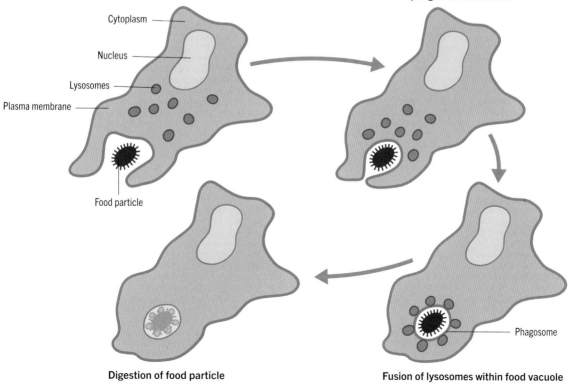

Entrapment of food particle

Cytoplasm

Nucleus

Lysosomes

Plasma membrane

Food particle

Formation of phagosome within cell

Phagosome

Fusion of lysosomes within food vacuole

Digestion of food particle

// Protozoa—movement

Using a microscope to watch an amoeboid cell clambering along on its pseudopodia is one of the more memorable experiences that a school biology class has to offer. These structures that it forms are certainly very reminiscent of the limbs of animals. Looking at all motile organisms, we see numerous examples of "walking legs," with independent evolutionary origins and built from a variety of materials, from jointed insect legs supported by a chitinous exoskeleton to octopus arms made of sinuously flexible muscular tissue, or the complex muscle-and-bone legs of vertebrates. Growing legs is clearly an evolutionary winning formula.

That pseudopodia can form (and, in some cases, maintain their shape over a long time) shows that the cytoplasm inside a protozoan cell is much more than a uniform jelly in its constitution—a runny jelly inside a flexible membranous bag would not readily hold a complex three-dimensional shape over time, even if supported by water. Cytoplasm does have structural elements though—various protein filaments and microtubules that collectively form what is termed the cytoskeleton. These can move around, pulling in sections of cell membrane here and expanding others there, to create and provide the necessary support to pseudopodia, without compromising flexibility.

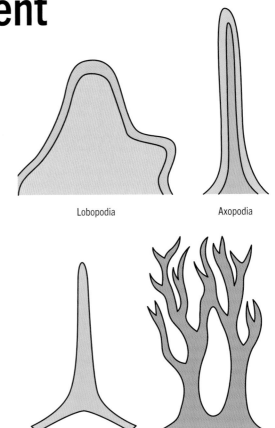

Lobopodia   Axopodia

Filopodia   Reticulopodia

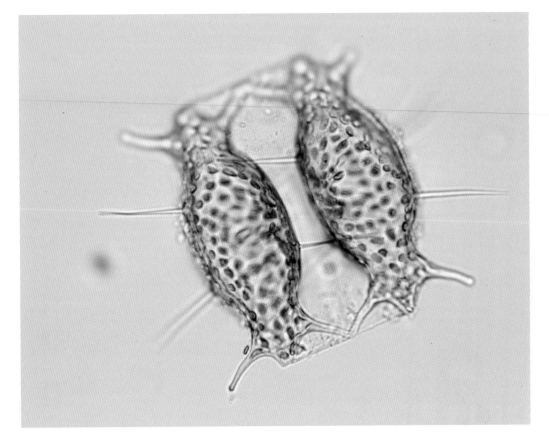

Above *Some of the different forms of pseudopodia.*

Left *The marine protozoa* Dinophysis *seen under the microscope. It is a genus of dinoflagellate found in a variety of coastal and ocean waters.*

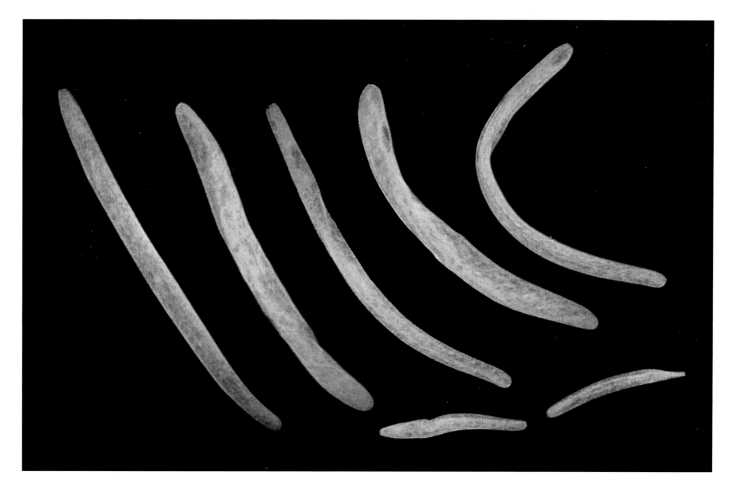

Above *The genus* Spirostomum *contains various long ciliates that are the fastest-swimming protozoa.*

Those protozoa that swim rather than walk use either flagella or cilia, which are (as we have seen) variations on the same essential structure and move with a swishing or whip-like action to provide propulsion. The ciliates are particularly swift swimmers and may traverse a microscope slide too quickly for you to get a clear look at them—adding a viscous material (prepared carefully so as not to harm the organisms through changes in osmotic pressure) to the liquid on the slide can slow them down to aid observation.

Some organisms usually regarded as protozoa do not have flagella, cilia, or pseudopodia. Among them are members of the group Apicomplexa, which includes the parasites that cause malaria and toxoplasmosis. These organisms can move, though, through successive adhesion and release of proteins on the cell membrane on a surface, creating a slow rolling or gliding motion.

HIGH-SPEED CILIATE

The cheetah, the peregrine falcon, the black mamba, the sailfish, and *Spirostomum ambiguum*. All five are exceptionally fast movers. The last-named, a slender pond-dwelling ciliate, may be the one fewest people have heard of, but it is also the fastest of them all, at least in terms of rate of acceleration. Its usual movement is done through the conventional beating of cilia, but it can also contract and extend its body to shoot forwards with an acceleration of up to 656 ft/s^2 (200 m/s^2). By contrast, a cheetah can only manage about 31 ft/s^2 (9.5 m/s^2), a peregrine barely better (and that is with the help of gravity, too).

// Protozoa—reproduction

Most protozoa reproduce by mitotic cell division (the process described on page 80). This transforms the parent cell into two daughter cells that are genetically identical, or at least extremely similar, to the original parent cell. It is a relatively swift and simple process in some cases, but some protozoa have more than one nucleus (multinucleate) and each of these has to divide as part of the process—in some cases this happens all at the same time but in others the nuclei divide at different times, making the total process more protracted.

Division by mitosis occurs more frequently when a population of protozoa finds itself in a favorable location with plenty of resources, for example when parasites arrive in a new host and begin to invade the host cells, or when free-living protozoa migrate to a more productive habitat. A protozoan cell needs to grow, through consuming extra organic material, before it can divide in this way, to ensure its daughter cells both have enough cytoplasm and other content to survive and thrive. Less favorable conditions favor sexual reproduction and spore formation instead (or the creation of a durable spore-like cyst), although not all protozoa are able to reproduce in these ways.

Budding, as seen in yeasts, is another way that some protozoa reproduce, with the new daughter cell containing the duplicated DNA in a newly formed nucleus but initially remaining attached to its "mother" as it grows. Sometimes the daughter grows within the mother cell rather than as an extension on the outside, and in multinucleate cells there may be multiple daughters developing on or within the mother. Budding remains an asexual reproduction method that generates genetic clones of the parent cell.

Some protozoan types can switch to sexual reproduction as required, and this may involve the production of gametes, or the coming together of compatible cells to exchange DNA. For others, however, a sexual stage is part of a complex life cycle, which involves a series of different cell forms. One example of the latter is the parasite *Plasmodium*, infective agent of malaria, which needs a mosquito host and a vertebrate host (say, a human) at different times in its lifecycle. When an infected mosquito bites a human, *Plasmodium* in a spore-like form enters the host and migrates to the liver, where it divides several times, resulting in multinucleate cells that enter the bloodstream and invade red blood cells. These rupture to release active cells that are capable of growth and asexual reproduction (and cause malarial symptoms). At this stage, some of the parasite cells produce gametes. When a mosquito now bites the infected human host and ingests these gametes, they combine inside the mosquito—i.e. sexual reproduction. The product of this is an oocyst, which ruptures to releases a new round of *Plasmodium* cells in their spore-like form, ready to infect a new human host when the mosquito takes its next bite.

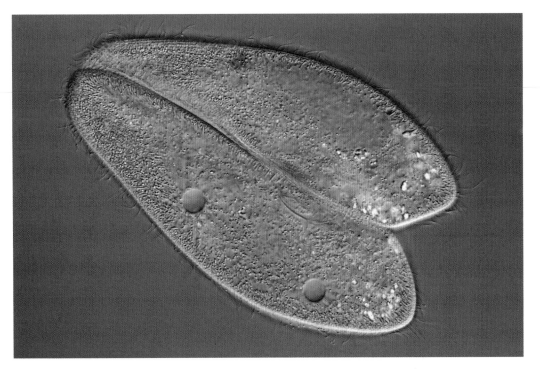

Left *Two* Paramecium *cells engaged in conjugation—a form of sexual reproduction unique to ciliates, which results in four "daughter" cells that share genes from the two parents.*

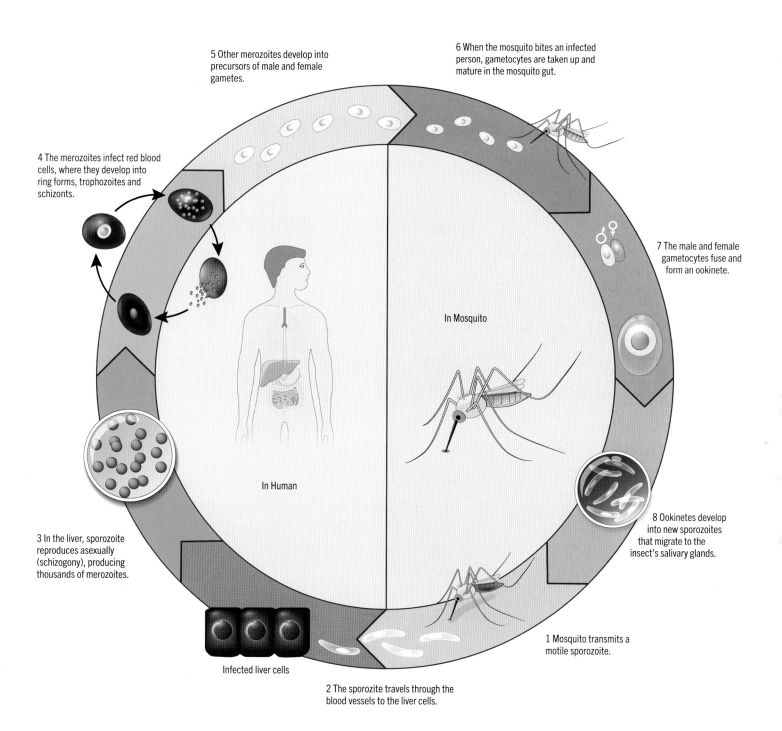

5 Other merozoites develop into precursors of male and female gametes.

6 When the mosquito bites an infected person, gametocytes are taken up and mature in the mosquito gut.

4 The merozoites infect red blood cells, where they develop into ring forms, trophozoites and schizonts.

7 The male and female gametocytes fuse and form an ookinete.

In Mosquito

In Human

3 In the liver, sporozoite reproduces asexually (schizogony), producing thousands of merozoites.

8 Ookinetes develop into new sporozoites that migrate to the insect's salivary glands.

Infected liver cells

1 Mosquito transmits a motile sporozoite.

2 The sporozite travels through the blood vessels to the liver cells.

Above *The life cycle of* Plasmodium, *the infective agent of malaria.*

// Protozoa as pathogens

Malaria is an especially prevalent and serious, often deadly, protozoan disease in humans. It is endemic in many countries and can only be held at bay by regular medication that kills the *Plasmodium* cells (it also helps to take whatever steps are possible to avoid mosquito bites). Interestingly, individuals with the inherited blood disorder sickle-cell anemia, in which the red blood cells develop in an aberrant shape, are more resistant to malaria because their red blood cells cannot hold large populations of the *Plasmodium* parasite (although sickle cell itself can be a debilitating condition).

Other protozoan diseases that affect humans include toxoplasmosis, which is common in soil and can also be caught through handling the droppings of infected mammals. As we have seen, toxoplasmosis infection can affect the behavior of mice, but there is also some evidence to suggest it can impact human behavior and cause lasting changes to the expression of personality traits. The actual illness toxoplasmosis is usually very mild, sometimes asymptomatic, and can be caught multiple times. However, if a pregnant woman becomes infected for the first time, the pathogen can seriously harm her unborn baby.

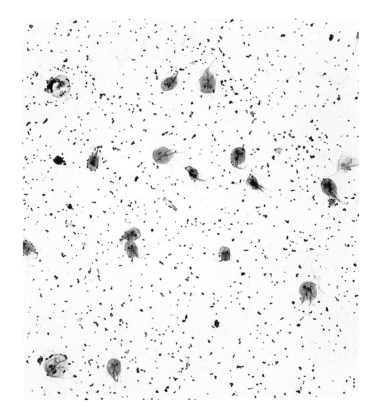

Above *The pathogen* Giardia, *cause of the disease giardiasis, which is common in humans and other mammals.*

Left *Infection with toxoplasmosis can cause a house mouse to lose its usually wary behavior.*

Giardiasis, caught by ingesting contaminated food or water, is spread via animal droppings and is caused by *Giardia*, a tiny flagellate that spreads by producing cysts that the host excretes. These cysts can remain viable for several months. It causes diarrhea and general intestinal distress, eventually causing weight loss and an inability to effectively absorb certain nutrients. Giardiasis affects a variety of mammals, including livestock and domestic pets, which may pass on their infection to their human owners. Diligent hygiene practices greatly reduce the chances of taking this parasite on board, and infected people do usually recover quickly as their immune system tackles the problem but medical treatment can be necessary (and prolonged) in individuals with weakened immunity.

Other protozoa that can cause disease in humans include the very common *Cryptosporidium*, which also causes gastrointestinal symptoms, and *Trichomonas vaginalis*, which causes a sexually transmitted disease. Protozoan disease in plants is much less common but can cause serious diseases and can additionally be a route for pathogenic viruses to move between plant hosts.

WATER AND DISEASE

Like other animals, we need to drink fresh water—we lose it constantly through our breath, perspiration and excretion, and we will die quite quickly if we don't replenish our stores. Even if we opt for tea, fizzy drinks, or fruit juice, it is the water that we really need. Water-borne pathogens have thus evolved a reliable way to move between hosts, which is one of the reasons why giving people access to clean drinking water is a healthcare priority around the world and is regarded as a fundamental human right. Of the world's 8 billion or so people, as many as 2 billion still have no regular access to safe drinking water, and water-borne pathogens are responsible for up to 80 per cent of the diseases that occur in developing countries. Treatment and transportation of water to meet the world's needs is a hugely costly process, and water security is a great concern for our planet's growing population.

Below *Provision of safe drinking water is not a given throughout the world. In many places people have to collect their water from untreated sources that may be brimming with harmful pathogens including salmonella and cholera.*

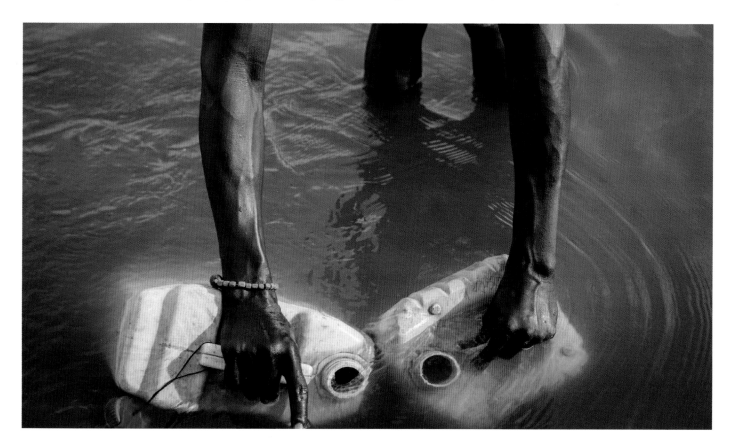

// Protozoa as symbionts

There is a fine line between a mutually beneficial relationship and a parasitic one (whereby one party exploits and may physically harm the other). Examples of both are replete throughout nature and the same association may veer either side of that fine line: take, for example, the oxpeckers—African birds that remove troublesome ticks from large grazing mammals, such as giraffes, but may also sometimes peck at healing wounds and drink their host's blood. The term symbiosis is usually used to describe a mutually beneficial relationship, but a more precise alternative term is mutualism—strictly speaking, symbiosis covers any organisms that live together in this way, regardless of who benefits and whether either party is disadvantaged.

Relationships of this nature are often especially complex when we look at organisms that live permanently on or inside other organisms. Nearly all taxonomic groups of microorganisms have some parasitic, pathogenic members and others that live symbiotically with a host—and some endosymbionts are absolutely vital to their host's survival. Protozoa and many other unicellular eukaryote groups sit on both sides of the symbiosis relationship, having endosymbionts of their own (in the form of long-ago engulfed bacteria that now function as mitochondria and plastids) as well as living as endosymbionts, and so they provide a dramatic illustration of the close interdependency of a wide variety of living things.

Below *The ciliate* Paramecium bursaria, *showing its numerous chloroplasts and cilia.*

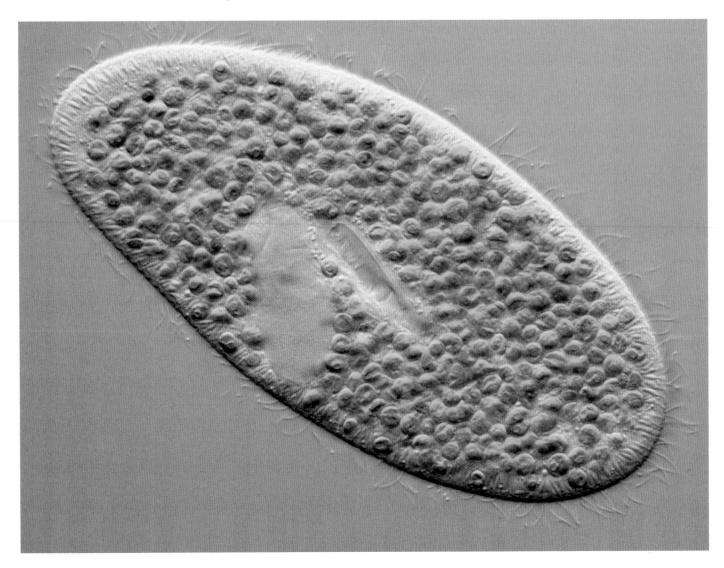

As we have seen, protozoa are important endosymbionts of herbivores, including termites and also cows and other ruminants, helping their host to digest a diet of cellulose from tough plant material that is indigestible to most plant-eating animals. Free-living protozoa, though, can have a beneficial impact on plant growth by consuming the soil bacteria that consume nutrients lost by the plant through its root system, and in the process releasing the nutrients and allowing the plant's roots to take them up again. The healthy human gut biome includes a variety of protozoan species alongside the more celebrated "friendly bacteria." Although their precise role within the biome is generally not yet well-studied, it is likely that at least some are beneficial.

Below *The ruminant digestive tract, with arrows indicating the food's direction of travel through the animal's four stomachs.*

PROTOZOA AS HOSTS OF ENDOSYMBIONTS

Paramecium, that much loved ciliate genus, includes one species (*P. bursaria*), which is host to an endosymbiotic green alga (genus *Zoochlorella*, some of which are endosymbionts with other organisms, such as sea anemones). The alga performs its usual photosynthesis, giving a supply of glucose and oxygen to its host, and benefits from protection and the *Paramecium*'s ability to quickly move around to the best light-gathering locations. However, cost–benefit analyses in varying conditions show that the deal generally is a much better one for the host than the endosymbiont, which would be better off without the association. A few protozoa species host bacteria as endosymbionts, the bacteria taking its share of nutrients from the protozoa while synthesizing certain compounds that the protozoa requires for growth.

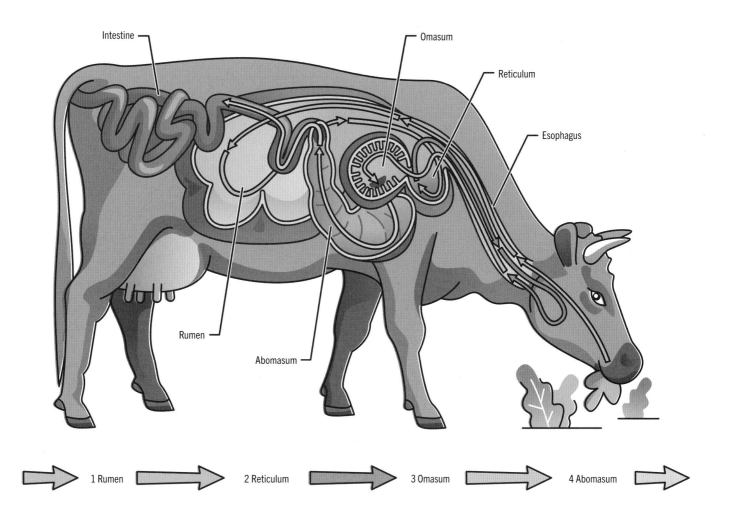

Intestine

Omasum

Reticulum

Esophagus

Rumen

Abomasum

1 Rumen → 2 Reticulum → 3 Omasum → 4 Abomasum →

// Protozoa—uses

Protozoa are just like us, in far more ways than we might like to admit. Consisting of one cell rather than many is a fairly big difference, admittedly, but our shared eukaryotic nature means that protozoa are valuable organisms for humans to study, from schoolchildren to biotech engineers. The use of organisms like *Paramecium* as protozoan "lab rats" has taught us a great deal about eukaryotic life in general, and expanded our views on how complex and beautifully adapted a "simple" life form can be.

Another important "model organism" within Protozoa is *Tetrahymena thermophila*. Studies on this easily reared ciliate have advanced our knowledge in a wide range of fields in microbiology, including the discovery of several organelle types and the first observations of various biochemical processes. *T. thermophila* has also proved to have an unusually complex mating system during the sexually reproductive phase of its life cycle, with seven distinct mating types or sexes, which can pair up and reproduce in 21 possible combinations. This species has even shown an apparent ability to learn. When switched from one swimming space to another, it will choose swim paths around the new space as if it were the same shape and size as the previous space at first, before becoming more adventurous and exploring the spatial limits of its new home.

As we have seen, protozoa have various roles in their ecosystems, and we can harness some of their abilities for our own benefit. Many free-living species are avid eaters of bacteria, and so are useful for cleaning up waste water in sewage works. Bacteria eat the organic matter in the waste water, and protozoa feed on the bacteria, resulting in cleaner effluent in sewage works with large populations of protozoa residing in their settlement tanks.

We have already seen how oomycetes may be used as biological pest control agents in agriculture and horticulture, to deal with plant-attacking fungi (see page 103). Similarly, parasitic protozoa can be used to control crop-damaging insects, as well as targeting insects such as mosquitoes that can spread serious human diseases, although this work is still in its infancy. Biological pest control is an important area of research, as (if done well) it would allow us to protect our crop yields without the use of highly toxic pesticides, which harm non-target organisms and can affect entire food chains.

Below Tetrahymena thermina, *a well-studied ciliate with a complex way of life.*

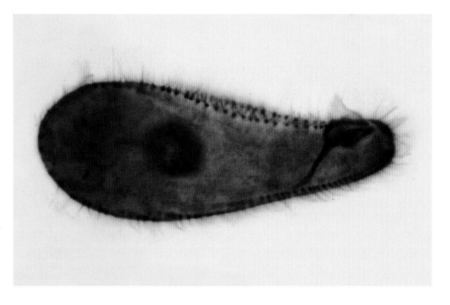

Below *Sewage treatment works rely on a community of micro-organisms to purify the water that passes through.*

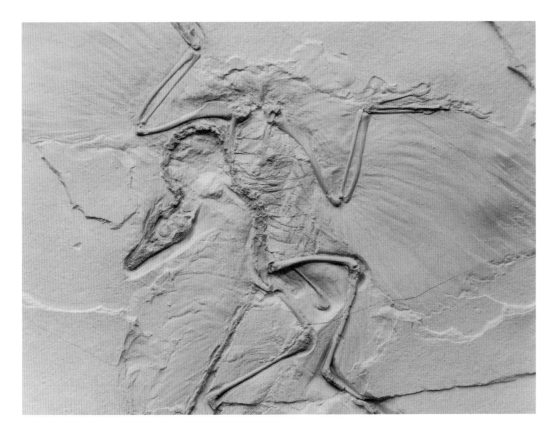

Left *Limestone is a rich source of fossils, including some, such as* Archaeopteryx, *that have proved pivotal in guiding our understanding of evolution.*

ROCK OF LIFE FORMS

A pebble or rock is the classic example of a natural but non-living thing, but some rocks owe their existence to long-dead unicellular life forms that, when alive, had an external shell. Limestone is a particularly important example of these "biogenic" rocks, and we use it in large quantities in building works, both as blocks and mixed with other substances to make materials such as cement and mortar. Areas with limestone rock formations are also happy hunting grounds for people who enjoy seeking out fossils—the famous and beautiful fossilized "first bird" *Archaeopteryx* was discovered in the Solnhofen Limestone formation in Germany.

// Macroscopic single-celled organisms

The only cell in the human body that we can see without the help of a microscope is a mature ovum (the female gamete), which is a sphere with a diameter of about 0.1 mm—the width of a single hair. The other 37 trillion cells are microscopic, averaging 100 micrometers across (and of course the 39 or so trillion prokaryotic cells that live on and in us are smaller still). So we conceptualize cells as tiny, and we are mostly right, but there are also a few impressively hefty unicellular organisms on our planet.

We have already met the biggest single-celled algal species on Earth. *Caulerpa taxifolia* is not only much bigger but also appears to be much more complex than many multicellular plants. Other *Caulerpa* species are similarly sizeable and have also convergently evolved a structure that is highly reminiscent of a true plant. Unsurprisingly, this remarkable and complex cell cannot reproduce by simple mitosis, but a ⅖ in (1 cm) fragment torn from it can grow into a full-sized organism, and it is also capable of sexual reproduction, producing tiny gametes that swim to find each other using flagella power.

Above *The remarkable and remarkably big single-celled alga* Valonia ventricosa.

Left *The trumpet-shaped alga* Stentor, *with its ciliated "mouth" busily gathering food particles.*

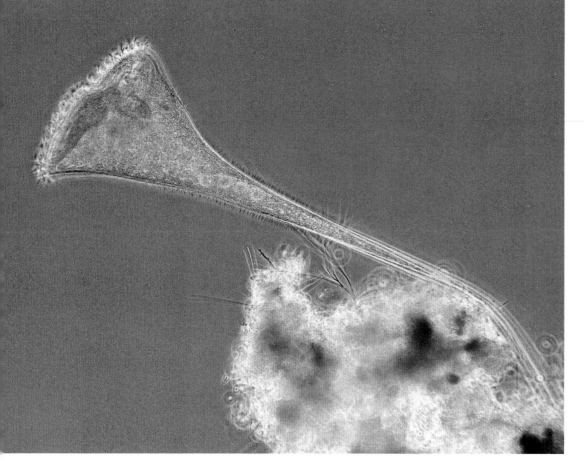

Above Caulerpa taxifolia, *native to tropical seas, forms a lush carpet on this Mediterranean seabed. Unfortunately, it is a harmful invasive species in this environment.*

The unicellular alga *Valonia ventricosa* is another oversized macroscopic species, but does not have an elaborate, differentiated structure. Instead, it recalls a green balloon or bubble—a glossy, taut-looking sphere up to 2 in (5 cm) across, sitting on coral rubble in shallow tropical seas. Inside, there is a large vacuole at the cell's center, and the other contents are divided into zones, each with its own nucleus and set of chloroplasts. It, like other *Valonia* species, reproduces asexually and sexually alternately, producing spores in the former phase and gametes in the latter.

The ciliated protozoa also include some supersized species, although they do not reach the proportions of the algae described above. The biggest are no more than ⅕ in (5 mm) long and include the deep-sea family Geleiidae, as well as the distinctive genus *Stentor*, which resemble tiny trumpets with their bell-shaped "mouths" into which a circle of cilia sweep food items. The biggest amoeba, *Gromia*

sphaerica, can reach 1½ in (38 mm) across. It is a shelled species and could be mistaken for a grape-sized pebble as it sits on a silty sea bed, but it is capable of movement and leaves tracks as it makes its way across the mud. A few types of deep-sea foraminifera grow to impressive sizes as well, including *Syringammina fragilissima*, which forms a shell up to 8 in (20 cm) across, with a complex but delicate ribbon-like structure. It was classified with the sponges (which are true animals, and multicellular to boot) until close examination proved it to be unicellular. These organisms are very little-known, thanks to their habitat and extreme fragility (the chances of moving specimens anywhere they could be easily studied, without damaging them, are minuscule). However, observations of their growth rate over eight months suggest that they could be exceptionally long-lived for single-celled organisms, as well as exceptionally large.

// Slime molds

Can there be a less alluringly named organism on Earth than the dog-vomit slime mold? This land-dwelling species, perhaps better known by its scientific name *Fuligo septica*, appears as a bright yellow, irregular blob (or perhaps a "splat" would be a better word) on the ground or on tree bark. It is one of the most frequently observed representatives of a highly enigmatic group of living things, the slime molds. They have long posed a headache for taxonomists because of their very strange appearance and behavior, and their uncertain relationships to other organisms.

For many years, slime molds were grouped within the kingdom Fungi, under the name Myxomycetes. Today, we know that the various species we call slime molds do not form a single group, despite their similarities, but that they belong to the group Amoebozoa, which also includes the more familiar amoeboid protozoa. The key trait that the slime molds share is the ability to switch between living as small single amoeboid cells and, under certain environmental conditions, forming large, continuous aggregations that behave like one organism. Most remain very small even in their aggregated form but some aggregations are very extensive and heavy. Some types form a single membrane-bound unit or plasmodium when they aggregate—effectively a huge single cell—these are known as acellular or plasmodial slime molds.

Above *With its dramatic yellow coloration and baffling physical form, the dog vomit slime mold is extremely striking.*

The other main group, the cellular slime molds, live as single cells until they are ready to reproduce, which is triggered by a shortage of food. Now, the individual cells release chemicals that others of their species recognize. Following these chemical trails, the amoeboid cells move together, forming an aggregation. This is a cluster of cells, each of which maintains its own membrane, but the cluster acts like one individual organism, moving slowly along like a tiny creature in its own right as it seeks out a suitable place to settle and form its spore-releasing structures.

Slime molds live on land, albeit in damp places, such as shady woodland floors. In rainforests, they may grow on tree branches and fruits high in the canopy. They consume organic material, both that which is already dead and decaying, and living bacteria and fungi. Depending on the species, they may reproduce asexually (by forming fruiting bodies, similar to mushrooms, which produce spores) and/or sexually (through production of motile gametes of as many as 11 mating strains).

Not all slime molds are terrestrial. The group Labyrinthulomycetes is ocean-dwelling, and when it aggregates some of the cells form a sort of mesh of tubular filaments, which are used as highways for cells to travel through the matrix. Some other species classed as slime molds are internal parasites of plants rather than free-living.

Above *Examined in close detail, the fruiting bodies of this* Lamproderma *slime mold are strangely beautiful.*

Opposite Tubifera ferruginosa *or red raspberry slime mold is a widespread and distinctive species.*

// Colonial algae

If predatory protozoa are like the tigers and leopards hunting alone in the jungle, then algae are like antelopes on the grasslands. The resource that they need to sustain themselves (in this case, sunlight rather than grass) is abundant and reliable. It does not need to be stalked, chased and overpowered—just be in the right place at the right time and you will have it in unlimited supply—and others of your kind will be drawn to those areas too. These are the sorts of conditions that can encourage group-living in animals, and a colonial habit in microorganisms.

Truly colonial algae live together in predictable group sizes, and arrange themselves in consistent ways. Members of the genus *Scenedesmus* are non-moving green algae that can live alone but also often form a small colony, or coenobium, of usually four or eight rod-shaped cells, joined along their sides in a tidy row. The colony is contained within a single cell wall. It may originate from a single parental cell that goes through a couple of rounds of cell division, but free-living mature cells may also coalesce into a group of eight. Colonies are more abundant than singletons when conditions are challenging—in warm and bright conditions, the cells appear to "prefer" to stay alone,

Above *Colonies of* Synura *algae gather in round, flower-like arrangements.*

but colonies are better at floating, keeping them closer to the sunlight. Being colonial also offers protection from algae-eating enemies, and improves survival chances when nutrients are scarce, so colonies are more likely to form where these hazards are abundant.

Below *As its name suggests,* Scenedesmus quadricauda *tends to form groups of four cells, lined up neatly in a palisade arrangement.*

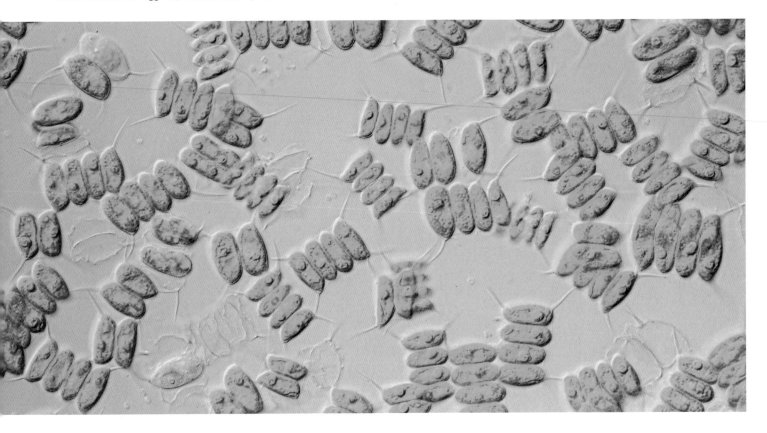

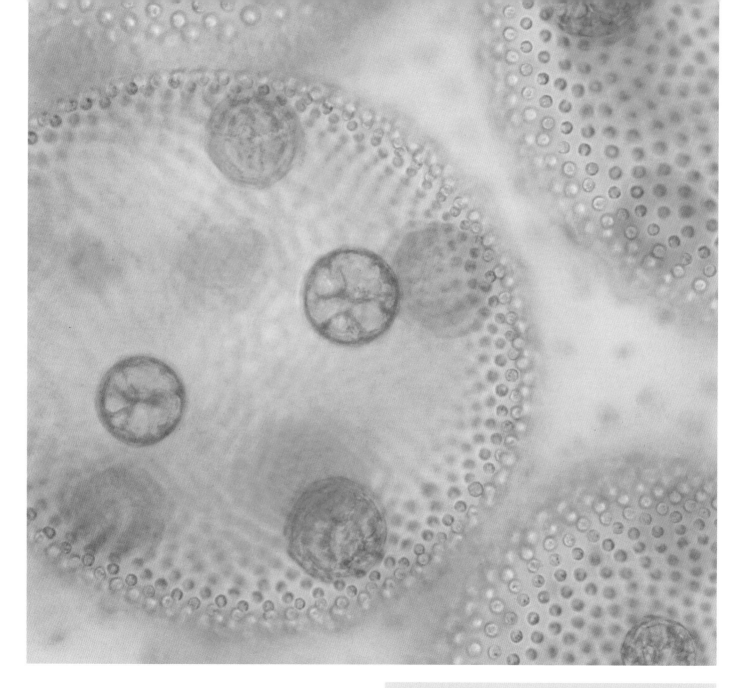

Above *A colony of* Volvox *algae in characteristic spherical form.*

The group Synurales includes various species of golden algae that have a scaly covering made from silicon-rich material. Those of the genus *Synura* form small, solid colonies that resemble ball-shaped flower heads, with each individual oriented with its paired flagella pointing outwards. The green alga genus *Gonium* forms colonies of up to 16 round cells that arrange themselves in a flat plate, stuck together with gelatinous material, and in terms of its evolutionary relationships as well as its colonial habit, it is intermediate between the genus *Chlamydomonas*, which exists as single cells (albeit often in large but non-connected aggregations), and the genus *Volvox*, which, as we saw on page 87, forms very large organized colonies that contain two distinct types of cells—a precursor to true multicellularity.

INSIDE OUT

All of the individual cells in the hollow, spherical colony of *Volvox* algae have their flagella pointing outwards, which allows them to swim along as a single entity. However, when a new colony is first formed inside the parental colony, the cells form the other way around, with the flagella on the inside. The sphere, therefore, needs to turn itself inside out before it can become properly functional. To accomplish this, some of the new cells have a wedge shape, which forces a bend in the spherical sheet, initiating the inversion process. That the colony can perform this contortion shows how elastic its structure can be, holding together as a sheet while it flips itself inside out.

// Other colonial protists

What is the difference between 1) a collection of cells that happen to be in the same place, and 2) a colony of cells? Or, to look at it from the other side, what is the difference between 2) a colony of cells, and 3) a simple multicellular organism? The answer is that these three states represent increasing degrees of interdependency and co-ordination, but there is perhaps no definite dividing line between 1 and 2, or 2 and 3. In theory, cells in a colony are not differentiated into different specialized types, and should be able to survive and take care of their own needs if they are separated from that colony. In practice, though, some colonial species are never seen in unicellular form.

What we can ascertain, from numerous examples across unrelated groups of organisms, is a definite evolutionary tendency for single-celled living things to form distinct, co-ordinated colonies under certain circumstances, sometimes involving multiple species, and to benefit from this arrangement. Sometimes the colony is temporary, as in the slime molds, with its members also having a unicellular life stage. In other cases it is permanent.

As explained, several types of unicellular algae habitually exist in colonies, which can be very large and also show the beginnings of cell type differentiation. The slime molds are rare examples of colonial protozoa, while members of the euglenid genus *Colacium* spend part of their life cycle in aggregations called palmellas, in which the individual cells are stuck together long-term by mucilage and are non-moving. Although yeasts are not generally regarded as colonial, when they grow hyphae, they are very similar structurally to molds, mildews, and rusts, which are considered to be colony-forming microscopic fungi.

MULTI-SPECIES COLONIES

In the world of non-microscopic, multicellular animals, a colony means something rather different. A group of grey herons who build their nests close together in the same tree have formed a colony. If a pair of little egrets come along and also nest nearby, then this has become a multi-species colony. Living in close proximity is beneficial to these birds in a range of ways, and in mixed-species colonies the different members may play different roles in colony life, with some being quicker to alert the whole colony to an approaching danger, and some being better at driving away predators. However, they may also compete and exploit one another. Microbes can also exist in mixed-species colonies and build similar positive and negative associations with one another—the communities that form the gut biomes of humans and other mammals are one example. For microbes, such situations also present opportunities for endosymbiosis to occur, which is arguably another kind of colonial life. A very important one, too, for as we have seen, all eukaryotes, from euglenids to elephants, came about because of endosymbiotic associations.

Opposite *In nature, both microscopic and macroscopic organisms may benefit from living in mixed-species colonies. Here, these black-headed herons, cattle egrets and reed cormorants benefit from a shared vigilance against danger.*

Below *The euglenid* Colacium *may be found in a sessile (non-moving) colonial form, or as single free-living cells.*

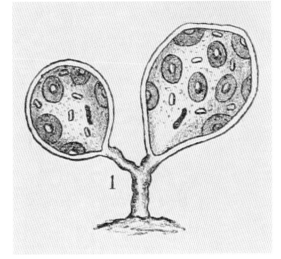

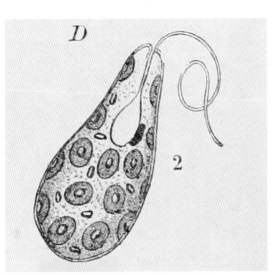

// Tissues

The trillions of cells in our bodies are differentiated into a couple of hundred different types, and populations of one or a few cell types form what we call bodily tissues. In some cases, multiple tissue types form distinct bodily organs. Higher plants and fungi also have differentiated cells that form distinct tissue types. We have now moved well beyond the colonial microbe stage of life's complexity, as the cells in species like these certainly could not survive and proliferate on their own. They are reliant on input of various kinds from many other cell types in the body in order to develop, survive and perform their particular functions, and collectively they form a set of co-ordinated systems.

Under the microscope, types of cells and tissues that you may have looked at in school include a very thin slice of onion, with its tightly stacked, rectangular cells, or a swab from inside a human cheek, containing plump, rounded squamous cells dislodged from the outer (epithelial) layer of the inside of the mouth. These two examples show some of the key differences between plant and animal cells (rigid cell walls versus pliant cell membranes) and how they tend to arrange themselves.

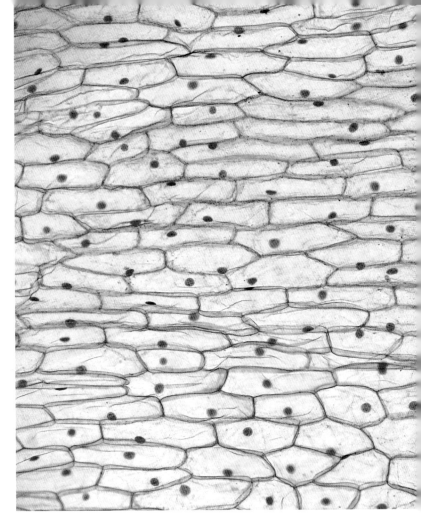

Above *Cells forming the epidermis of an onion.*

Opposite *Squamous cells from inside a human mouth, which together form epithelial tissue.*

Not all multicellular organisms have distinct tissues. Some of the simplest animals, the group Placozoa, comprise a ball of cells that moves in an amoeboid motion—its cellular differentiation is rather minimal, and it can reproduce via budding of a part of the parent animal, similar to many unicellular organisms. However, it does have distinct cell types to handle different functions, for example ciliated cells to sense its surrounds and to provide movement, and food-digesting cells to secrete digestive enzymes into trapped food particles. Among plants and their close relatives, seaweeds in the brown algae and green algae groups have distinct parts, which may include a stipe (stalk), a holdfast (similar to roots), and leaf-like fronds, although as we have seen, the *Caulerpa* algae have these parts too, all as part of just one cell. These multicellular algae do have a varied cell population forming distinct tissue types, but they are considerably less specialized and differentiated than those we see in true plants.

Above *A multicellular brown alga (*Fucus*) with differentiated parts (including bladders to help it stay afloat).*

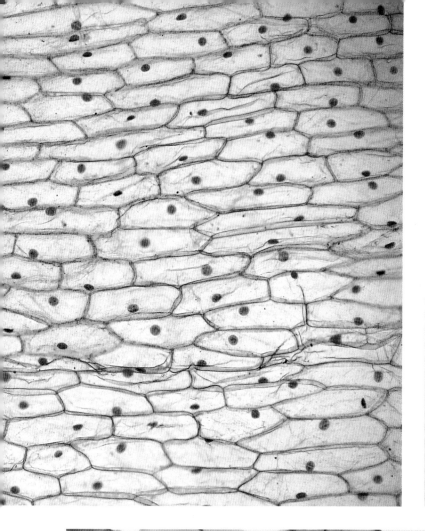

THE PROS AND CONS OF COMPLEXITY

From our perspective, being a multicellular organism with numerous different bodily systems all working together can feel like a wonderful miracle. The way of life that this complex body grants us certainly seems a lot more potent and rewarding than the existence of a protozoan swimming about in a ditch. We (and many of our fellow animals) can move in so many different ways, form enduring and multi-layered relationships with one another, experience emotion, heal from injury, and explore and survive in a variety of habitats. Yet complexity also brings some particular vulnerabilities. The most significant is a loss of adaptability. While we may be quite adaptable as individuals, as species we have too slow a reproductive rate and too slow a mutation rate (a driving factor of evolution) to adapt to rapid environmental change. These are challenges that unicellular organisms (and especially prokaryotes) are much better equipped to handle. If through our own actions we make our planet a hostile environment for ourselves, we can be pretty sure that single-celled life will fare much better than we do.

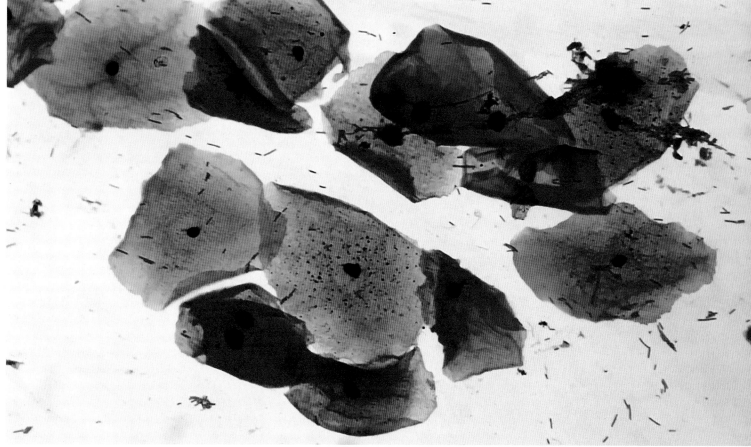

MICROANATOMY OF TISSUES AND STRUCTURES

Cells of the same type, existing in aggregations, have the potential to share their resources in various ways. When they become truly interdependent, then what was a colony of cells becomes a multicellular organism. Most of the living things that we are familiar with are multicellular eukaryotes. Some of these living things are still technically microscopic, but all of them, from the tallest tree to the hugest whale, are made up of a multitude of microscopic cells. These cells exist in true and total interdependency, which has allowed them to become differentiated and specialized into a range of types, to carry out one particular task within the body as a whole.

The cells in our bodies are differentiated into many different types, forming tissues such as bone.

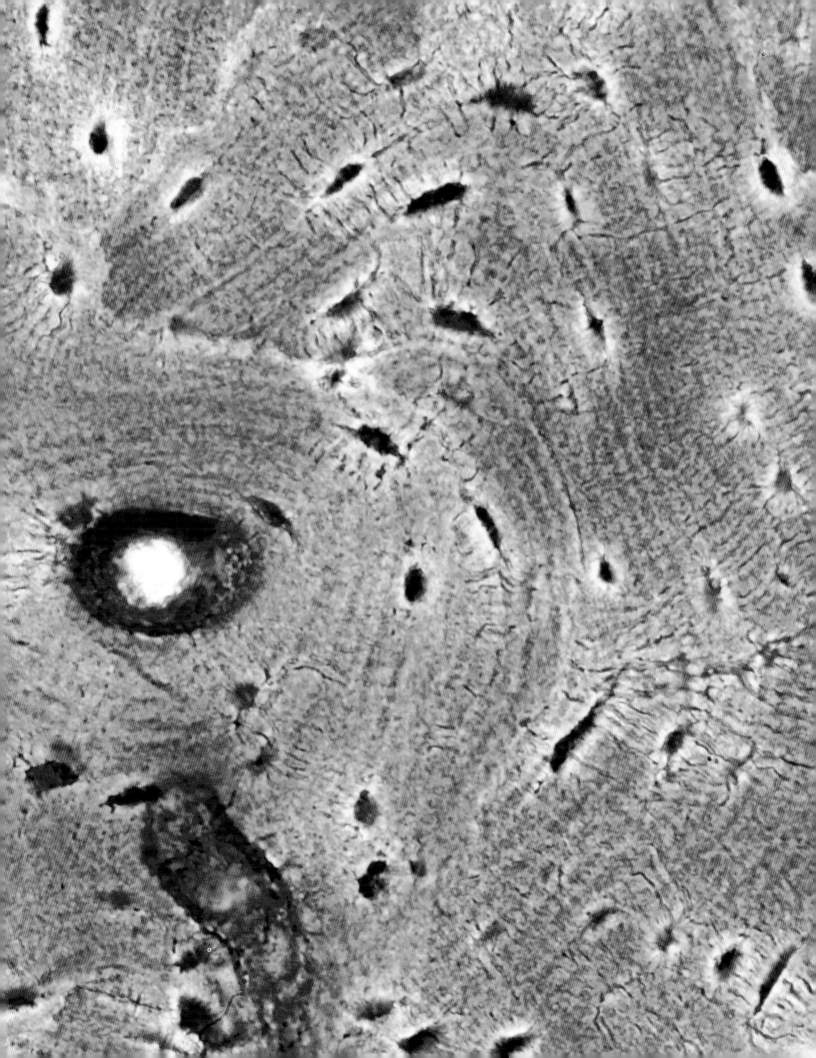

// Types of multicellular eukaryotes

The multicellular organisms on Earth are hugely outnumbered by single-celled life, but from our perspective they are disproportionately significant. Most of these organisms (and certainly the ones we are most likely to see, know, and study) belong to one of the three "classic" eukaryote kingdoms—Plantae, Fungi, or Animalia. Their bodies contain several differentiated types of cells that carry out distinct functions, and while some are truly microscopic, in most cases you don't need a microscope to visualize them clearly (unless it is the cells themselves that you want to see).

The plants are distinguished from multicellular algae in that they have a system for moving water and other substances around from cell to cell, whereas in algae the individual cells can take up and release water as needed. In most plants, this is comprised of two specialized tissue types (xylem and phloem) that form vessels to carry nutrients, water, and waste from one part of the plant to another. These tissues enable a 330 ft (100 m)-tall tree to carry nutrients and water up from the tips of its roots (via xylem), and transport the glucose made during photosynthesis downwards from the leaves to the roots (via phloem). As they are non-motile but often reproduce sexually, some flowering plants have a mutualistic relationship with animals (especially insects), which move their male gametes to the ova-bearing parts of other flowers, in exchange for a nectar reward.

Fungi, more animal-like than plant-like in their cell structure and feeding habits, grow through their thread-like

Below *The roundworm* Caenorhabditis elegans, *a relatively simple and very small animal.*

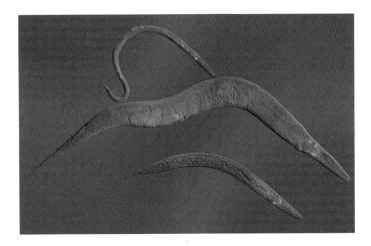

hyphae, producing a root-like network called a mycelium, through which they absorb nutrients. At certain times, they produce fruiting bodies to disperse their spores. These visible parts of a fungus are often very striking to us, but by far the larger part of these organisms is the mass of mycelia growing out of sight, underground. It is often entwined with the roots of living plants in a symbiotic association, helping the plants extend their reach and absorb more water and nutrients. The fungus benefits by consuming carbohydrates made by the plant. This "mycorrhizal network" also serves as a system for communication and nutrient exchange between different individual plants.

Unlike plants and fungi, which are mostly land-dwelling, the majority of animal species live in water, both fresh and marine. Most animals are motile, some extremely so, which means that on land they have important ecological roles as dispersers for plants and fungi. Some tiny animals have a set number of cells in their bodies—a phenomenon called eutely. For example, the male nematode worm *Caenorhabditis elegans* always has 1,033 cells, while its hermaphrodite form (which can self-fertilize) has 959 cells—but these cells are organized and differentiated into many types, to form a distinct nervous system, reproductive system, digestive tract, musculature and sensory organs. The simplest animals in terms of variety of cell types are the placozoa (see page 128), which can have as few as six different cell types.

500 μm

Above *A microscopic view of fungal hyphae connected to a rice plant's root.*

Opposite *Pint-sized representatives of the three most familiar Eukaryote groups—a snail (animal), mushroom (fungi) and the fruiting bodies of a moss (plant).*

// Microscopic plants

The simplest plants are not necessarily the smallest, at least at first glance. Some members of the group formerly known as bryophytes or "lower plants," which comprises mosses, liverworts and hornworts, can grow extensively in suitable conditions. A continuous blanket of moss, though, is composed of individual plants that may be just 0.08 in (2 mm) tall (although the biggest, members of the genus *Dawsonia*, can reach 20 in/ 50 cm tall). These three groups of simple mini-plants have much in common in terms of general appearance and ecology, but may not be close relations in evolutionary terms.

Mosses and their relatives lack true roots but do have leaves (or at least leaf-like structures, lacking the complex structure of "higher" plant leaves), which are green with chloroplasts and grow with a wide surface area to capture sunlight. Under a strong hand lens or microscope, the incredibly delicate structure of moss leaves is revealed—some are just one cell thick—but they are also detailed and intricate, growing as slim fronds that bear overlapping pointed teeth or rounded lobes. Liverworts can be leafy or grow thicker-looking, lobe-like tissue. Hornworts look similar to lobe-like liverworts.

When bryophytes reproduce sexually, the mature plant develops tiny male gamete-producing structures (antheridia) and female equivalents (archegonia, in which a single female gamete develops). Male gametes, which have flagella and are able to swim through a water film, move to archegonia where fertilization occurs, the two haploid gametes uniting to form a single cell with a full genome—a diploid zygote. A stalked, encapsulated structure called a sporophyte forms around

Below Dawsonia, *an exceptionally large moss species. For most mosses, a hand lens is useful to properly examine the structure.*

the new zygote—in some mosses, you can often see these projecting well above the mossy foliage, as little blobs on slender stems. When mature, the sporophyte capsule opens (at this stage, a magnified view shows that the opened-up capsule now looks like a tiny, delicate flower) and releases spores. These (provided that they land in suitable conditions) will germinate and develop into new plants. Bryophytes can also spread through asexual reproduction—a broken-off piece of leaf can regrow as a new plant, a genetic clone of its "parent."

THE TINIEST FLOWER

Dip your hand in water containing *Wolffia arrhiza* and when you bring it out it will be covered in what look like little yellow-green ovoids, each about a millimeter across. You might think that this is an alga, but in fact you are holding lots of individuals of the world's smallest flowering plant, as well as the smallest vascular plant. It is therefore a member of the angiosperm lineage—the most complex of plant types on Earth—despite its diminutive size. Known as spotless watermeal, this plant lives in fresh water and, although it usually reproduces vegetatively by budding, it can also reproduce sexually via the microscopic flowers it forms, which sit inside a depression in each plant. It is a useful species to us, being capable of purifying waste water and (when growing in clean water) being a useful food source.

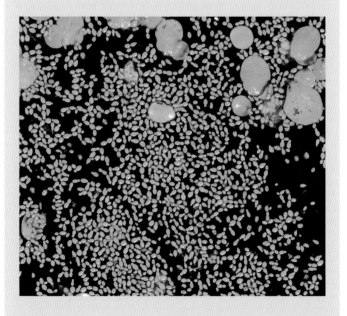

Above *A true flowering plant, but in miniature—the species* Wolffia arrhiza.

Below *The sporophyte capsule of a moss, from where the reproductive spores are released.*

// Plant tissue types

The plant cell that comes to mind most readily for most of us belongs inside a leaf. It is boxy in shape, its cell wall pressed snugly up against its neighbors on all sides, and it is full of vivid green chloroplasts, within which photosynthesis is taking place throughout the daylight hours. In leaves, cells like this form a tissue called parenchyma, which sits just under the leaf's epidermis. Parenchyma is a type of ground tissue. Internal plant parts are mainly made of ground tissue, with dermal tissue forming outer layers, and vascular tissue forming the vascular system.

Right *This cross-section through a plant stem shows the various tissue types, with dense and fibrous supportive sclerenchyma on the outside and looser parenchyma tissue inside.*

Below *Sclereid cells within the tissue of a pear fruit.*

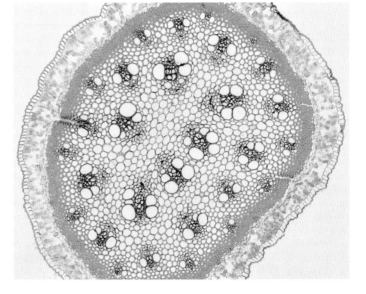

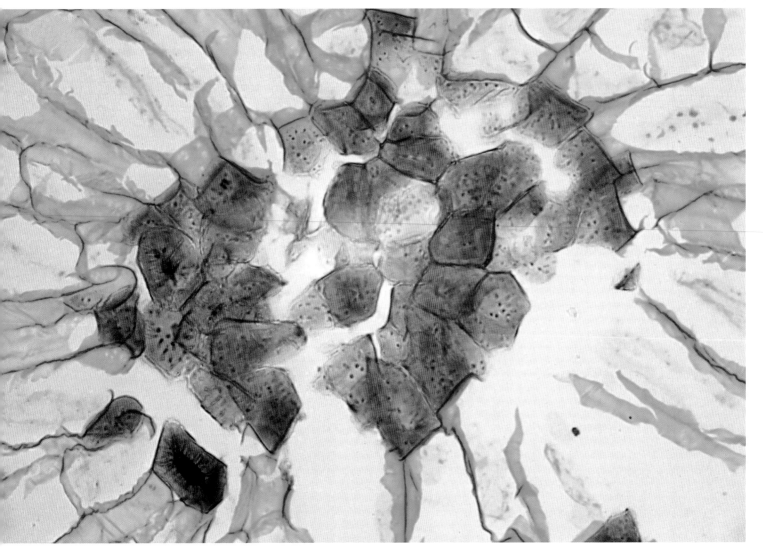

The epidermis covers all of a plant's leaves (on both top and bottom), stems, and roots, and is made of dermal tissue—in this case, a single layer of flat cells. It includes paired, horseshoe-shaped guard cells that surround the openings (called stomata) where air can enter and leave. The epidermis has a protective, sometimes waxy outer coating of lipids and hydrocarbons, called a cuticle. Under the epidermis, the parenchyma tissue includes palisade mesophyll and spongy mesophyll cells, which are respectively tall and tightly packed, and rounder and more loosely arranged, with spaces to allow the free movement of the gases that the cells take in and release. These two types of mesophyll cells are where most photosynthesis occurs. Elsewhere in the plant, parenchyma is formed of rather undifferentiated cells and carries out many active functions as well as photosynthesis.

Plants have two other kinds of ground tissues. There is sclerenchyma tissue, which (once fully developed) is not living. Its mature, dead cells are thick-walled, robust, and rigid, and function as support structures to give the plant stability. Tree bark and outer wood is a type of sclerenchyma, which serves to protect, insulate, and support the living tissues deeper inside. The constituents of sclerenchyma include very long cells called fibres, which provide support in stems, boughs, and trunks as well as smaller structures, and the more varied sclereid cells, which sometimes originate from parenchyma and form structures like the shells of nuts, as well as being scattered in softer plant parts to aid support.

The other ground tissue type we see in plants is collenchyma. This is a sturdy tissue of thick-walled cells, found as a living support element in growing shoots and leaves. The xylem and phloem vessels are formed from bundles of vascular or transport tissue. They contain elongated conduction cells that actually carry out the transportation of nutrients (these cells are tracheids in xylem, and sieve cells in phloem). The conduction cells are physically supported by sclerenchyma and are assisted in some of their functions by parenchyma cells.

Plant tissue containing cells that are not yet differentiated is called meristem. Its cells are small, thin-walled with large nuclei, and divide regularly, becoming more differentiated and gradually developing into a mature tissue type as they go through repeated divisions. The epidermis, ground tissues, and vascular tissues all arise from different forms of meristem—respectively they are: protoderm; ground meristem; and procambium.

Below *Cross-section through a leaf, showing different tissue types.*

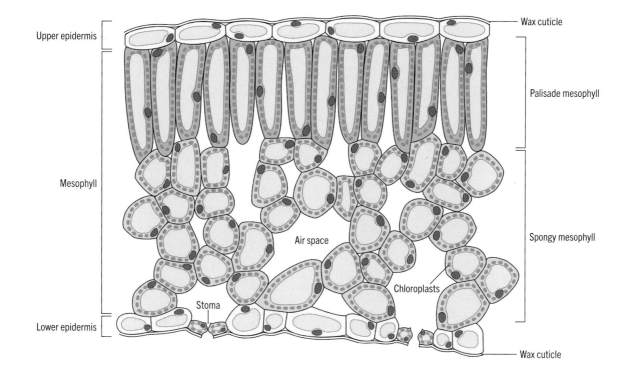

// General plant structures

A typical plant is held in the soil by its branching root system, which provides water and anchorage. Above ground, it grows a stem that often branches to reach upwards and/or outwards, and on its terminal stems it grows leaves that are shaped to capture sunlight for photosynthesis, the product of which will feed the entire plant. With some plants (usually smaller ones) there is no main stem, just leaves. However, reproductive structures are usually borne on stems, whether these be flowers, cones, or spore-releasing structures.

Above *Our green planet owes its color to the chlorophyll in plants, carried in plastids that were once microscopic cyanobacteria.*

Left *A typical plant has an above-ground part which carries out photosynthesis, and a below-ground part for collecting water and nutrients from the soil.*

A plant's roots are not exposed to the light (except in the case of aerial roots) so their cells do not require chloroplasts for sunlight-gathering and photosynthesis. They do, however, have gathering of their own to do, of water and vital minerals dissolved in that water, so their root cells grow fine, delicate hair-like projections to maximize their reach into the surrounding soil. Leguminous plants also grow round nodules on their roots, which are homes for nitrogen-fixing bacteria. Swollen below-ground structures such as tubers, rhizomes, and corms, found in various flowering plants, are used for food storage (and a food source for humans). Bulbs also store food, allowing a plant to regrow after its foliage has died back in winter. These structures may seem to have a uniform constituency when we cut them up, but like other plant parts, they contain different cell types. A potato is mainly made of parenchyma cells, full of stored starch, and it also has a vascular system. The skin is made of three kinds of dermal cells, and immediately below it is a layer of procambium.

Plant stems may be permanent and woody, with large quantities of the structural polymer lignin in the cell walls of the supporting tissues. Stems of plants that die back in winter (herbaceous plants) have less lignin and are softer. Some woody plants that live in temperate climates grow and shed a full set of leaves every year, while others, including most tropical species, retain leaves for longer, although they do shed older leaves and constantly grow new ones. Leaves come in a wide variety of shapes and growth patterns, from long and narrow to rounded, lobed, and pennate (where numerous small leaflets grow either side of a central stem). Their purpose is the same in all cases—to present a wide photosynthesizing surface to the sun during daylight hours— so they tend to grow in such a way as to not shade each other.

HOW PLANTS SENSE LIGHT

Plants have no eyes (by our understanding of what an eye is) and yet they clearly sense light. You can manipulate the growth of a windowsill houseplant by rotating it—the leaves naturally grow towards the light source. They can do this because of a variety of protein molecules in the cells that are activated in the presence of light, in some cases to particular light wavelengths. These photoreceptor molecules not only guide the movement of living leaves, but also play a role in triggering plant growth in response to day length changes.

Above *Compounds within plant leaf cells respond to light and collectively they can cause the leaf's orientation to shift around, to capture more of that radiant energy.*

Right *Plant cells contain various photoreceptors, which are sensitive to different light wavelengths.*

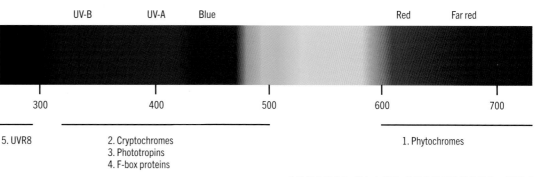

// Respiration in plants

As we have seen, respiration is a crucial metabolic process in living things. It involves the breakdown of glucose molecules to create a molecule called ATP, which the organism uses as a form of chemical energy. In terms of gas exchange, aerobic respiration is the opposite to photosynthesis (it consumes oxygen and releases carbon dioxide), so it's important to remember that, just like animals, plants do require oxygen and do release carbon dioxide, as well as being providers of the first and consumers of the second.

Typically, the green parts of plants carry out most of their respiration in the hours of darkness, when photosynthesis has ceased, but their root cells will respire at all times. The by-products of this are released into the surrounding soil, which benefits the community of microorganisms that live around the root system, and this is a symbiotic association because the microbes break down organic matter into a form that the plant's roots can take up.

The breakdown of one molecule of glucose uses six molecules of oxygen and generates six molecules each of carbon dioxide and water. The uptake and release of these molecules takes place via the stomata—holes in the epidermis, most noticeably on the undersides of the leaves. The guard cells that sit in pairs on either side of each stoma can change shape to make the stoma smaller or larger—a wider opening allows gases to move more quickly in and out of the leaf, and a smaller one helps to prevent too much water loss through evaporation—some of the water formed in respiration may be needed elsewhere in the plant. The reaction of respiration generates heat within the mitochondria, which in a few plant species can be considerable, and beneficial. The main advantage is that it heats up volatile compounds that the plant produces to attract its pollinators. The skunk cabbage, which grows in cold regions, can use its heat to melt a snow covering so that it can photosynthesize.

Above *This skunk cabbage produces enough heat (through the activity of its mitochondria) to melt away its snow covering and give it access to vital sunlight.*

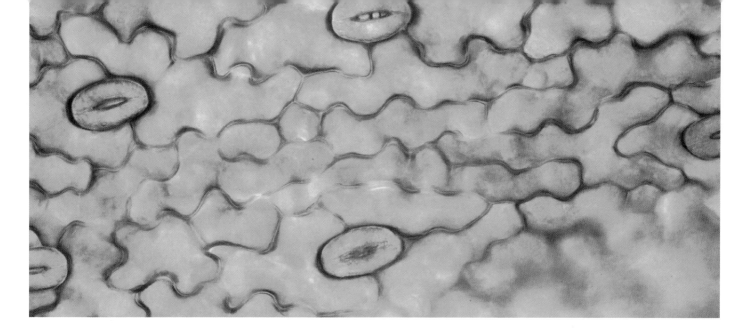

Above *Plant epidermis on a leaf. The oval-shaped pairs of guard cells alter the size of the openings between them (stomata), controlling intake and loss of gases.*

Respiration requires the presence of mitochondria, which are as ubiquitous in plant cells as they are in animal cells. These structures are visible under the microscope as sausage-shaped organelles, looking similar in both cell types, and are believed to have the same origin as endosymbiotic bacteria (and to descend from an event that pre-dates the acquisition of chloroplasts in the ancestors of plants). However, plant mitochondria contain a much larger genome than animal mitochondria, with between 10x and 100x the number of base pairs. Much of this additional material is

non-coding, though, so the actual amount of coding DNA is comparable between the two. Plant mitochondrial genomes are also much more prone to change through generations as the mitochondria duplicate themselves, with sections of DNA frequently being lost, gained, duplicated, and rearranged. The DNA in plant mitochondria also exists mainly as separate strands, rather than a single, closed-ring molecule as in animal DNA, which would explain why plant mitochondrial DNA goes through more changes, as the opportunities for genetic "mix-ups" during replication are higher.

Below *Structure of a typical plant cell. Its rigidity is provided by a sturdy cell wall, and a large interior vacuole full of fluid.*

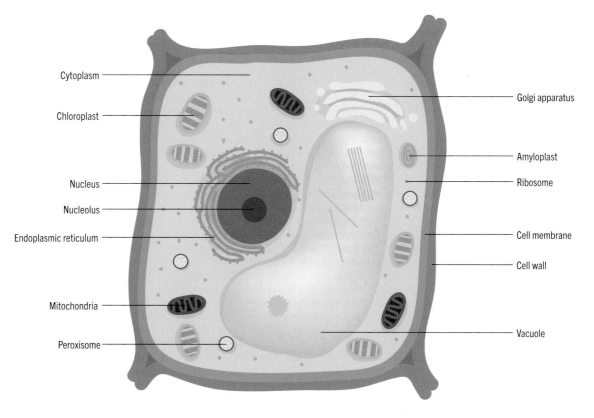

Cytoplasm

Chloroplast

Nucleus

Nucleolus

Endoplasmic reticulum

Mitochondria

Peroxisome

Golgi apparatus

Amyloplast

Ribosome

Cell membrane

Cell wall

Vacuole

// Photosynthesis in plants

If you seal yourself in an airtight chamber, the size of an average household room, after about three days you'll be close to death because of insufficient oxygen and (more importantly) too much carbon dioxide. An atmospheric CO_2 concentration of anything over 5 percent is dangerous to us, damaging cells and interfering with our blood's ability to absorb whatever oxygen is still available. Given that our exhaled breath is 4.4 percent CO_2, it won't take too much breathing to turn the chamber's atmosphere into one in which you can't survive. However, if you cram about 700 carefully chosen houseplants in there with you, and keep the lights switched on so that your leafy companions can photosynthesize non-stop, you'll live much longer—long enough for other issues to become more of an immediate threat, certainly.

We are well aware that photosynthesis provides atmospheric oxygen and removes carbon dioxide, and therefore that we need there to be ongoing photosynthesis if we are to live. We tend to vastly underestimate the contribution of all of those microscopic, marine photosynthesizers to this process, which is much greater than that of the true plants, but both are needed to keep our planet's population of animals and other consumers alive.

We also depend completely on the food created through photosynthesis, whether we eat the photosynthesizers themselves or obtain the nutrients via an animal intermediary. Glucose is made in the chloroplasts and then used or converted into other molecules elsewhere in the cell.

As we know, chloroplasts, site of photosynthesis, descend from free-living cyanobacteria. They are 1–10 micrometers across on average, some are round and others are more ovoid, and a typical mesophyll cell in a leaf contains between 20 and 100 of them. When you look at plant cells under the microscope, the chloroplasts are the most obvious organelles, thanks to their green color. Like mitochondria, they retain a genome (with genes that code for their various types of RNA) and replicate on their own, independently of the parental cell. The large central vacuole found in many plant cells, which stores water and maintains the cell's shape, also pushes the chloroplasts to the outside of the cell, where they have more opportunity to receive sunlight.

Gas	Inhaled air	Exhaled air
Oxygen	21%	16%
Carbon dioxide	0.04%	4%
Nitrogen	78%	78%

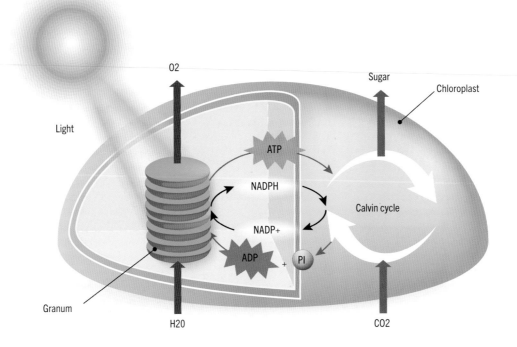

Above *The complex process of photosynthesis depends on two compounds: nicotinamide adenine dinucleotide phosphate hydrogen (NADPH) and adenosine triphosphate (ATP). These are required to drive its final stage, the Calvin cycle, whereby glucose (sugar) molecules are built.*

Above *The parasitic plant dodder, entwined around and penetrating its host's stems to intercept supplies of food.*

PARASITIC PLANTS

Dodder, a plant of the genus *Cuscuta,* goes by various unflattering alternative English names—strangleweed, hellbine, beggarweed, pull-down. They allude to this curious, pale yellowish plant's parasitic ways—rather than photosynthesizing its own food supply, it winds itself around a host plant (which it locates and grows towards by detecting chemicals released by the host), invades its tissues, and steals a share of the food that host plant makes through its own photosynthesis. It does this by growing a root-like structure called a haustorium, which penetrates the host's vascular tissue and intercepts nutrients moving through the host. A single dodder plant can grow multiple haustoria and parasitize several host plants at once. About 1 percent of flowering plant species are partially or fully parasitic.

// Metabolism in plants

Glucose, which plants build when they photosynthesize, is a simple sugar molecule, consisting of a chain of carbon atoms bonded to hydrogen and oxygen atoms in (typically) a ring-shaped configuration. It is quite reactive in this form and can be chemically combined with other glucose molecules and other types of compounds. All of the proteins, fats, and structural and food-store carbohydrates in a plant are built using glucose molecules. When the plant requires energy, its food-store (starch) molecules must be broken down again into their constituent glucose molecules. These are then broken down themselves, during respiration, to provide energy. All of these activities constitute the plant's metabolism.

Metabolic processes come in two general categories—catabolism is the breaking down of molecules, and anabolism is the building or synthesis of new molecules. A plant cell contains different organelles for handling different aspects of metabolism, with chloroplasts and mitochondria being key components of that system as the former creates glucose and the latter breaks glucose down for energy release. Mitochondria are also the site of amino acid synthesis, and plant cells assemble all 21 of these themselves (in animals, several need to come from the diet as they cannot be made by the cells). The DNA in the cell nucleus holds the code for how amino acids need to be arranged to synthesize all of the proteins the plant uses, and its code is carried out of the nucleus via molecules of messenger RNA. The actual assembly of amino acids into proteins is carried out by ribosomes. These proteins include enzymes, which play important roles in assisting (catalyzing) a range of metabolic reactions.

Starches are built in both chloroplasts and another type of plastid, found in root cells, called amyloplasts. Cellulose, the other main glucose polymer molecule found in plants, is synthesized in a moving region of the cell membrane called the cellulose synthase complex, from where it moves directly to the cell wall. Fatty acids are built in plastids as well and are then assembled into long-chain fat molecules in the endoplasmic reticulum.

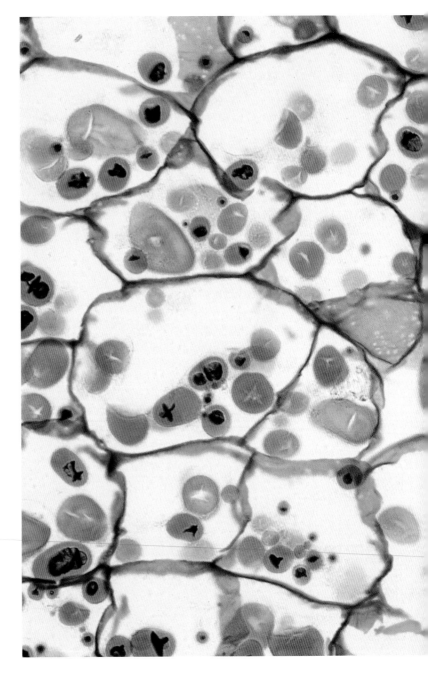

Below *Some plant-derived enzymes that have industrial uses.*

Enzymes	Source	Application
Bromelain (cocktail of proteases)	Pineapple stem	Anti-inflammatory agents, meat tenderizer
Papain (protease)	Papaya latex	Anti-inflammatory agents
Urease	Jack bean	Determination of blood urea nitrogen
b-amylase	Barley	Brewing industry
Ficin	Fig latex	Food industry
Peroxidase	Horseradish	Diagnostic industry
b-glucanase	Malted barley	Brewing industry

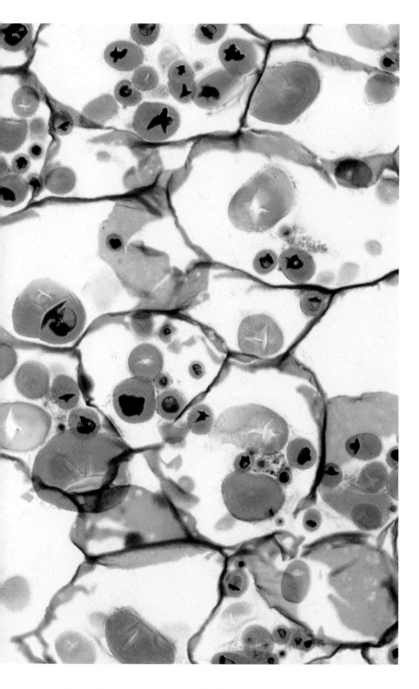

PLANT HORMONES

We tend to associate our own hormones with shifts in our moods and behavior, so it may seem strange to us that plants, which we consider entirely unemotional and mostly inactive, are as highly hormonal as we are. Hormones have many roles in both plants and animals, though. These chemical messengers, which may be fat-based or protein-based, control rate and timing of growth and development, storage of nutrients, division and differentiation of cells, and the yearly cycle of flowering and dying back again. They are built in very low quantities in many different cells (rather than in specialized organs, as in animals) and move through the plant through passive diffusion between cells (passing through tiny pores or plasmodesmata in the cell walls) or in the plant's vascular system. Their release may be triggered by signals—for example, various growth hormones are released when photoreceptor molecules respond to increasingly long hours of daylight as spring approaches. Unusual plant development, such as leaf-like structures growing where flower petals should be, may be down to a hormonal problem.

Above *These cells, taken from the flesh of a potato, contain starch-storing plastids called amyloplasts.*

Above *The distortion of this daisy flower, or fasciation, is the result of a hormonal problem.*

// Reproduction in plants

We often choose the gift of a gorgeous bunch of flowers to express passionate love, and well we might, as flowers are the organs of sexual reproduction in most of the world's land plants. Not all flower species would look (or smell) appealing in a vase on the table, but all have the same function—to produce male and/or female gametes that, when they meet, form the beginnings of new plants. Asexual reproduction methods are also very widespread in the plant kingdom, with many plants able to grow from a small broken-off piece.

A typical flower that is pollinated by insects or other animals will carry male and female parts on the same flower structure. The male parts, the stamens, produce pollen, a fine granular material composed of male gametophytes (see below). Female gametophytes develop in ovules that are part of the flower's pistil. The pistil has a long thread-like style, tipped with a sticky bulb called a stigma. The flower's shape is such that a visiting pollinator comes into contact with both anthers and stigma, and as it moves from flower to flower, it constantly collects pollen on its body from anthers, and transfers it to stigmas. Some other plants rely on the wind to carry their pollen to other flowers.

Under a powerful microscope, pollen grains appear as balls or other symmetrical shapes and show a complex, often ornate and spiky surface. They need to be tough to survive their journey, but once attached to a stigma, they extend a pollen tube into it, and two male gametes travel down this tube into the pistil. Meanwhile, the female gametophyte, developing inside an ovule, matures into one gamete, plus six other cells. Once one of the male gametes unites with the female gamete, the ovule then develops into a seed, which can germinate as long as it finds its way to a suitable area of soil.

SPOROPHYTE AND GAMETOPHYTE

Plants alternate their generations between a non-sexual phase (sporophyte) and sexual phase (gametophyte). As soon as it is formed as a single cell, from the fusion of two haploid gametes, a plant is a sporophyte. It grows now through cell division, and in due course produces spores, which are genetic clones of itself. In flowering plants and other "higher" groups, the spores remain inside the plant's male and female structures and mature through further cell division to form haploid gametophytes. Now, pollination occurs, during which the male gametophytes unite with female gametophytes, forming a new sporophyte and beginning the cycle over again. "Lower" plants, such as mosses and ferns, release their spores, and as long as the spores land on wet ground, they too will produce gametophytes. Male gametophytes of these types of plants are motile and can swim to find female gametophytes, in order to combine and create new sporophyte plants.

Below *Some examples of pollen grains.*

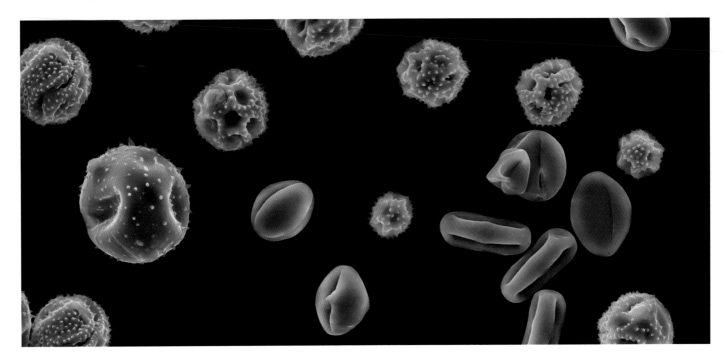

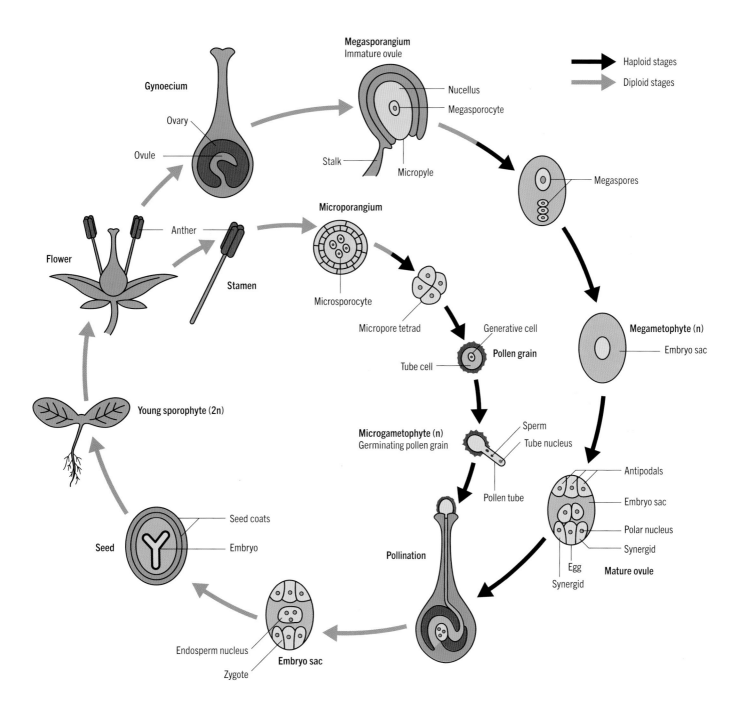

Gynoecium
Ovary
Ovule

Megasporangium
Immature ovule
Nucellus
Megasporocyte
Stalk
Micropyle

Megaspores

Haploid stages
Diploid stages

Flower
Anther
Stamen

Microporangium
Microsporocyte
Micropore tetrad

Generative cell
Pollen grain
Tube cell

Megametophyte (n)
Embryo sac

Young sporophyte (2n)

Microgametophyte (n)
Germinating pollen grain

Sperm
Tube nucleus
Pollen tube

Antipodals
Embryo sac
Polar nucleus
Synergid
Egg
Synergid
Mature ovule

Seed coats
Seed
Embryo

Pollination

Endosperm nucleus
Embryo sac
Zygote

Above *The sporophyte and gametophyte stages in a flowering plant, including the process of gamete maturation and fertilization.*

// Germination and growth in plants

The growth of a ⅕ oz (5 g) acorn into an oak tree standing 78 ft (24 m) tall and weighing more than 15.4 US tons (14 tonnes) certainly demonstrates how very able plants are to turn light, water, and soil nutrients into living tissue. The tree might take more than 200 years to attain that size, but we can see a seed becoming a mature plant much more quickly if we pick another species—plant a pinhead-sized poppy seed, for example, and you'll be admiring its pretty scarlet flower just 60 days later. This may not be a more impressive feat than the development of a human being from fertilized ovum to new-born baby in nine months, but the seed manages its own growth, and that growth is a lot easier for us to observe.

Below *An oak sapling's first few leaves are as large and as efficient as the 200,000 or more leaves it will bear when fully grown.*

By the time a seed is in a situation where it is ready to germinate, the zygote inside has already been through multiple rounds of cell division and many of these cells have differentiated into distinct tissue types—there is effectively a tiny but well-formed plant concealed under the tough coating (testa). As we have seen, in flowering plants, two male gametes enter an ovule at a time, and while one of these fertilizes a female gamete to form a diploid zygote, the other male gamete unites with two of the other six cells that have formed from the female gametophyte. This forms a second cell within the ovule, and while the zygote develops into the embryonic plant, this second cell is destined to become endosperm—a food store for the developing embryo. Its cells store large amounts of starch and protein as it grows within the parent plant's pistil, and by the time the seed is ready to germinate, the endosperm takes up most of its interior space.

A fully developed seed usually has a testa and its interior is mostly filled with endosperm. Water can pass in and out via a tiny pore (micropyle), and the embryonic plant (or cotyledon) itself, comprised of the epicotyl (leafy part) and radicle (root) is already partly formed close to the micropyle, the latter eventually emerging through this opening as it grows.

Above *Growth of a plant from seed.*

Right *Reproductive parts of a typical flower, and the process of fertilization.*

Some plants grow from seed to a particular size, while others continue to grow continually, gradually increasing in height, width, number of leaves, and total biomass year after year. Growth occurs through mitotic cell division and differentiation of meristem tissues, elongating and branching off stems and roots and broadening leaves. The process is under the control of various hormones, which take their cues from receptor molecules that respond to light levels as well as from information about available supplies of water and nutrients. In temperate and polar regions, low winter light levels drastically reduce plants' ability to photosynthesize and thus gather resources. Many cease to grow at all this time, while others die back so that resources can be conserved to support the plant's underground parts.

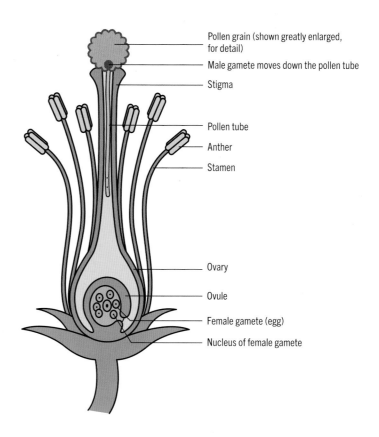

Pollen grain (shown greatly enlarged, for detail)

Male gamete moves down the pollen tube

Stigma

Pollen tube

Anther

Stamen

Ovary

Ovule

Female gamete (egg)

Nucleus of female gamete

// Color and pattern in plants

Most plants photosynthesize, and their living, above-ground tissues are therefore green because this is the light wavelength reflected by the photosynthetic pigment chlorophyll. Plants also contain other pigments in their cells, such as carotenoids and flavonoids, which are yellow or orange, and anthocyanins, which produce red-violet tones. These pigments modify the shade of green we see, and in some cases, their color can mask the chlorophyll enough to completely change the appearance of leaves and stems. The ornamental copper variant of the common beech tree is one example, its deep red-violet foliage produced by high concentrations of anthocyanins. Some ornamental plants have variegated leaves, with patterns in green and white. This is usually the result of a mutation affecting chlorophyll distribution. Having redder-colored or variegated leaves can reduce the plant's photosynthesizing efficiency, so these variations are rarely seen in wild plants.

Below *Types of plastids, and their development from precursors*

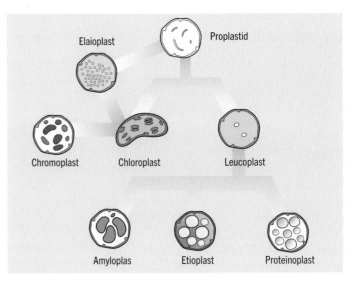

Left *How a buttercup looks to us (left) and to a bee (right). Note the ultraviolet reflectance at the bases of the petals, showing the bee where to access nectar.*

Below left *Flowers pollinated by birds tend to be red, and hummingbird feeders make use of this association.*

Flower petals are modified forms of leaves in terms of their tissue types and growth pattern. Their cells generally have less chlorophyll than leaves, but other pigments are more pronounced, so they dramatically outshine leaves when it comes to color and pattern. Those vivid yellows, blues, reds, and other shades, and striking patterns of streaks, splashes and spots, have evolved for visual appeal. However, we are not their target market, and we do not necessarily see them the same way that the plant "intends." Bees and other day-flying, nectar-eating insects, the chief pollinators of flowering plants, can see reflected ultraviolet light, so what we see as a relatively muted petal pattern can look much more strongly contrasting to their eyes. Often, the patterns on petals serve as a guide, directing the pollinator to where it will find nectar (and come into contact with the flower's anthers and stigma on the way). Those flowers that are pollinated primarily by birds (such as hummingbirds) tend to have red petals, a color that is less visible to bees and so does not attract them as much. Night pollinators like moths and bats see contrast much more clearly than color, so the flowers they pollinate often have white flowers, as well as strong perfume to tap into their pollinators' acute sense of smell.

The organelles that carry non-chlorophyll pigments in petal cells are chromoplasts. These plastids both synthesize and store pigments, and in terms of development they are derived from chloroplasts, which are present in the petal's cells in the flower's early development. Some plastids remain unpigmented—these leucoplasts give rise to white petals. They are also present in root cells and play a role in storing starch and other nutrients.

Below *The vivid colors of autumn leaves are always present, but it is only when the chlorophyll is lost that these other shades are revealed.*

AUTUMN LEAVES

The changing colors of deciduous tree leaves in autumn provoke wistful feelings, their beauty a reminder of the coming winter. Those colors, though, have been there all along. The loss of chlorophyll is what reveals the other cell pigments to us. Within the leaf's chloroplasts, the chlorophyll molecules break down and are replaced constantly through the lifespan of the leaf, but in autumn the renewal rate slows down, the tree redirecting those resources elsewhere as it prepares for winter survival. In due course, the leaf no longer makes enough food through photosynthesis to sustain itself, and so it dies and detaches from the tree.

// Predatory plant structures

Of the four elements that are most abundant in the constituents of plant cells, three (carbon, hydrogen and oxygen) are derived from water and air. The fourth, nitrogen, presents more of a challenge. Our planet's atmosphere has N_2 in abundance, but (with the exception of a couple of yeast species) eukaryotes are unable to access atmospheric nitrogen, so must obtain it in a different form—nitrates, which are compounds in which nitrogen is combined with hydrogen and oxygen. Animals get their nitrates from the plants and other animal they eat but, as producers, plants need another way. They can take up nitrates from broken-down organic material in the soil, through their roots, and a few have symbiotic associations with prokaryotes that can obtain (fix) atmospheric nitrogen and convert it into a useable form. Then there are a few plant species that approach nitrogen acquisition in the same way as animals—by eating it.

Above *A pitcher plant. The fly is safe as long as it does not venture inside.*

Left *The surface of a Venus flytrap's trap, showing the hairs which, when stimulated in the right way, trigger the trap to close.*

The Venus flytrap is probably the best-known carnivorous plant. Its traps consist of paired, modified leaves with a scarlet interior and fringes of long hair. A trap sits open until an insect alights on it. This triggers the two halves to close, imprisoning their victim, and the lengthy digestion process begins once the trap is fully closed and digestive enzymes can be released into the interior space. The trap's movement is produced by the two halves rapidly switching shape from inwardly to outwardly curved. It is triggered by stimulation of tiny, pressure-sensing hairs, formed from modified cells that are especially rich in energy-generating mitochondria, on the trap's inner surfaces. Two separate trigger hairs must be touched for the trap to close (so it doesn't snap in response to a non-living object landing on it), and another five stimuli from the struggling prey are needed to close it completely. The Venus flytrap's ability to count is less celebrated but no less remarkable than its ability to generate much faster movement than we would expect of any plant.

Other carnivorous plants include sundews, which trap insects on beads of sticky secretions, produced by specialized glands on the tips of fine hairs that cover their modified, tentacle-like leaves. The tentacle curls around the prey, sometimes startlingly quickly, to ensure it is held fast. Pitcher plants have large bucket-like leaves with steep slippery walls, and digestive fluids at the bottom of the bucket—like many carnivorous plants, they produce chemicals attractive to insects.

Above *The fly-trapping "glue" at the tips of a sundew's hairs is secreted by tiny glands—the secretion also carries digestive enzymes.*

JASMONIC ACID

Release of digestive enzymes in a Venus flytrap is triggered by the release of jasmonic acid, a hormone synthesized in the chloroplasts. It is found in many other species and has an important role in plant self-defense systems. Plants no more want to be eaten than animals do. They cannot run away, but they can fight back, and jasmonic acid is part of that process. When part of a plant is damaged by a munching herbivorous insect, a biochemical pathway is activated that results in the release of jasmonic acid. The hormone stimulates cells around the injury to release chemicals that block the insect's protein-digesting enzymes, thus discouraging it from continuing to eat the plant.

// Plant communication

When we want to urgently communicate with a nearby person, perhaps to warn them of imminent danger, we shout and wave. The importance of communication in our species is so great that we have, as a global culture, devised swathes of technological solutions to our need to make meaningful contact with people thousands of miles away. Plants, lacking voices, movement, and satellites, are nonetheless proving to be remarkable short-range and long-range communicators, using various biochemical and electrical signals to "talk" not only to one another but to (and through) other plant species and various animals and fungi as well.

Chemicals used as airborne signals by plants are known as VOCs (volatile organic compounds). Plants of many species use them to attract their pollinators and to deter potential enemies. They also warn one another about nearby hazards. For example, when an insect begins to eat one of them, that victim releases VOCs, which neighboring plants sense via receptor molecules in their cell membranes. Very quickly, the neighboring plants respond by increasing the amount of toxic or repellent chemicals they produce in their foliage, to defend themselves against the herbivore. They may also release VOCs that attract parasitoids—insects that lay their eggs on the early stages of other insects—as another line of defense. As we have seen (see page 143), the parasitic plant dodder has adapted to exploit this channel of communication between its host plants, sensing their VOCs and growing towards them.

Communication via the roots enables plants to signal to neighbors of drought conditions and other problems, allowing the neighboring plants to take action—for example, a plant can reduce its water loss by reducing the size of the stoma openings in its leaves. A plant's roots also allow it to communicate with other individuals over greater distances via the mycelia of fungi that connect them together—the mycorrhizal network (see page 132). At a microscopic level, the plants and fungi associate intimately, with the hyphae surrounding and growing into the cells of the plant's roots and forming a dense branched mass of hyphae (an arbuscle) within some cells. This allows easy and direct exchange of nutrients and water between the plant and the fungus, and also allows both participants to respond appropriately in the event of environmental changes such as drought.

Below *When a caterpillar bites into a plant leaf, the plant releases chemical "distress signals," which may attract parasitoid wasps to attack the attacker.*

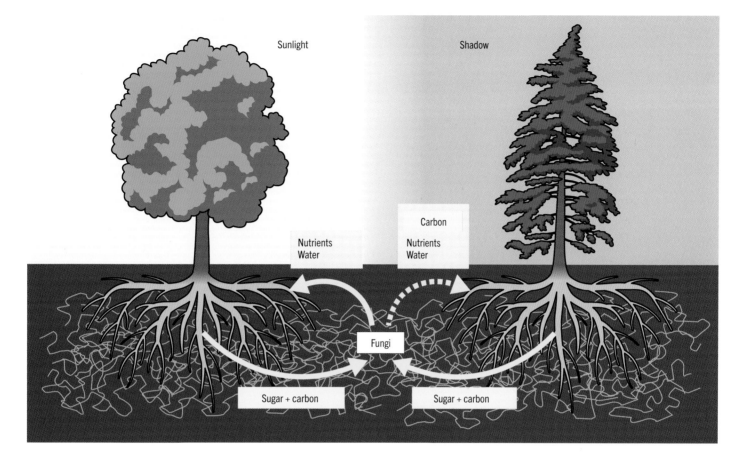

Sunlight

Shadow

Carbon

Nutrients
Water

Nutrients
Water

Fungi

Sugar + carbon

Sugar + carbon

ELECTRICAL SIGNALLING

Cell membranes have a different electrical potential (amount of energy needed to conduct electrical charge) on the outside compared to the inside. This is maintained by the concentration of positively and negatively charged particles (ions), which can be moved through the membrane via special channels. Membrane electric potential gives the cell a way to transmit an electrical signal to other cells. It is especially important in the nervous system tissues of animals, transmitting nerve impulses through the body, but plant cells are also able to communicate with each other in this way, much more quickly than is possible than chemical signaling.

Above *How trees and fungi share resources through the mycorrhizal network.*

Right *Recent studies show that tobacco plants and some other species make high-pitched sounds when they are damaged or dehydrated. The "calls" vary depending on the type of stress they are experiencing.*

// Microscopic fungi

The smallest organisms classed as fungi are the unicellular yeasts, which we met on page 96. There is also a range of fungi which, while multicellular, release their spores from extremely small fruiting bodies, and they are classed as microfungi. Among them are several pathogenic organisms, affecting animals and plants. Many cause mold and mildew on organic material, including items of perishable foodstuffs that we leave in our fridge a little too long, and can colonize damp places in our homes. Some species are involved in symbiotic association with lichens, while many are free-living soil organisms.

As with other fungi, the growth components of microfungi are hyphae. These extremely thin threads grow through the substrate, branching regularly, and collectively they form the fungal mycelia. An individual hypha may be just one cell wide. The cells from which it is formed have outer cell walls that are relatively robust, although water and nutrient molecules pass through them. In most species, the walls that separate each cell from its neighbor are somewhat different to the outer walls. These internal walls are called septa, and contain larger pores, big enough for some cell organelles to move through. Growth occurs with new cells being assembled at the hypha tip, and growth direction may be guided by environmental cues (as well as diverted by obstacles that the hyphal tips cannot penetrate).

Below *Aspergillus fungus, which can cause the lung disease aspergillosis.*

Although the fruiting bodies of microfungi are tiny and you will need a good lens (at least) to see them well, they are just as varied, strange, and attractive as the much more obvious mushrooms and toadstools sprouted by the so-called macrofungi. Some microfungi, though, simply release their spores from ordinary cells rather than growing distinct fruiting bodies. The mold that is often found on the peel of slightly elderly oranges (a *Penicillium* species—see below) produces tiny but elaborate fruiting bodies that you will not easily see with the naked eye—but you can't miss the cloud of dusty spores that float away from the mold if you pick up the orange.

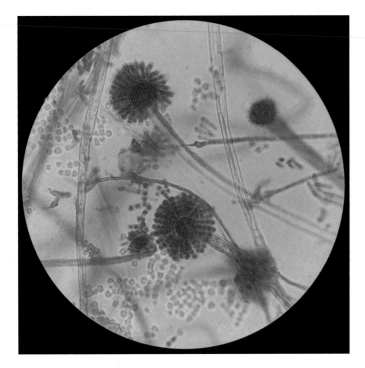

Right *The tiny fiber-like stuctures on this mold are reproductive fruiting bodies, releasing microscopic spores.*

Below *The distinctive blue-grey mold that grows on old oranges is caused by a fungus of the genus Penicillium.*

SICKNESS AND CURE

People having antibiotic treatment for the bacterial infection tuberculosis sometimes come down with another illness just as they are recovering. Aspergillosis is a mold-causing microfungus, which can also cause a very unpleasant lung disease. It is also rare, but with reduced competition from bacteria thanks to the antibiotics' action, it can take hold. Other pathogenic microfungi cause conditions ranging from athlete's foot to ringworm. However, probably the best known microfungus is the genus *Penicillium*, source of the highly effective antibiotic that (almost) shares its name. This microfungus's ability to kill off cultures of bacteria was discovered in 1928 by Scottish microbiologist Sir Alexander Fleming, and his discovery revolutionized how we treat a range of serious diseases. His work also revealed how rapidly bacteria can develop antibiotic resistance when an antibiotic dose is too low.

Above *Lab equipment set up to test the effectiveness of the antibiotic streptomycin, first isolated in the 1940s, on pulmonary tuberculosis.*

// Fungal tissue types

The cells and tissues that make up the various parts of fungi are mostly not as well differentiated as those of plants or animals. However, there are several reasonably distinct fungal tissue types, although not all may be found within any given species or specimen. Some of these relate to distinct hyphae in varyingly close arrangements, but other cell types are only found in fruiting bodies.

Hyphal tissue in general is known as plectenchyma. It has a natural tendency to clump together, because when the growing tip of a hypha contacts another hypha, they will often fuse together. When the hyphae grow in a tight bundle, but remain distinct strands, they are known as prosenchyma, but if they become fused together and homogenous, this is pseudoparenchyma (borrowing its name from the similar tissue type found in plants). Long-standing, tough and compacted clumps of hyphae are known as schlerotia—it is from these regions that fruiting bodies are most likely to grow. When hyphae form an elongated, root-like structure, this is a rhizomorph.

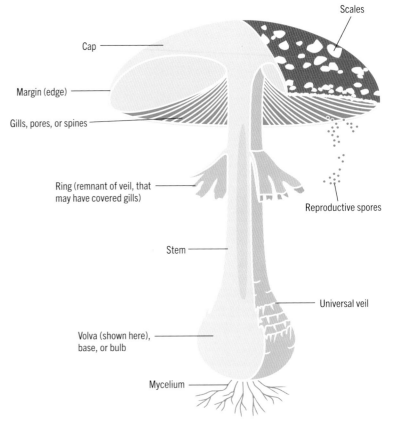

Above *Structure of a typical basidomycete fruiting body.*

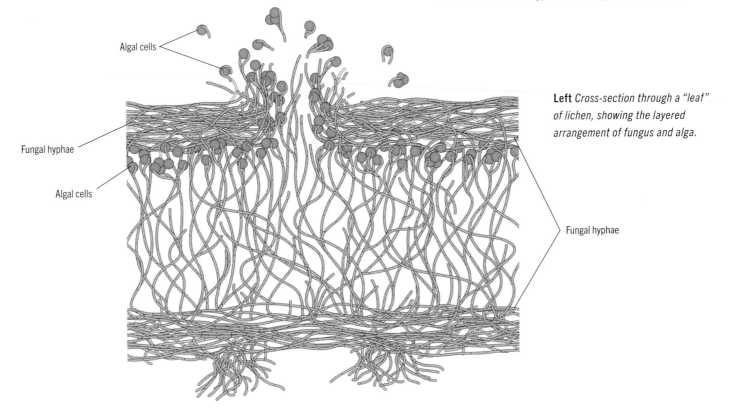

Left *Cross-section through a "leaf" of lichen, showing the layered arrangement of fungus and alga.*

Above *The spores released by a mushroom originate between the folds of its gills.*

In simpler fungi, which include many members of the group Ascomycota, spores are released directly from structures in individual hypha called sporangia, which develop in specialized hyphae. The reproductive spore cells they produce may be motile, swimming with the use of one or a pair of flagella, or immobile. More evolutionarily advanced fungi, belonging to the other major fungal group, Basidiomycota, grow elevated fruiting bodies, or basidiocarps, for spore dispersal. The main body of a basidiocarp is formed from pseudoparenchyma, supported by specialized, sturdy variants of fibrous hyphal tissue. The gills on the underside of the cap carry spore-producing cells (basidia).

The tissues of lichen form a special case. As we have seen, these are composite organisms, comprising an alga or a cyanobacteria living in symbiosis with a fungus. In firmer lichens, the two components may be divided into distinct layers, which are visible when you look at a magnified piece of lichen in cross-section. In jelly-like lichens, both organisms are mixed and woven homogenously together as a rather jelly-like mass.

EDIBLE FUNGI

The standard mushrooms we might buy in the supermarket to eat (*Agaricus bisporus*) have dense, homogenous white flesh in their caps, and more fibrous, hypha-like tissues in their stems. These fruiting bodies are usually harvested well before they mature, though, especially in the case of "button mushrooms"—in nature the caps would flatten out with growth, exposing the gills, which would become darker and larger, ready to shed spores. More mature *A. bisporus* are sold as "portobello mushrooms." Other mushrooms used as food may be quite different in texture—*Hericium erinaceus* or lion's mane, for example, has loose and stringy flesh.

Above *Edible mushrooms are often rich in B vitamins.*

// Metabolism in fungi

Like animals, fungi need to consume organic material, which they convert into the molecules they need to provide energy, maintain their cells, and assemble new ones as they grow. Lacking mouths, guts, or anything else resembling a digestive tract, they break down their food externally (extracellular digestion), by secreting molecules of their digestive enzymes directly into the food. Metabolism costs energy, which the fungus restocks through respiration, either aerobic or anaerobic.

The enzymes go to work chopping up large molecules into smaller ones, eventually breaking them down into molecules small enough for them to be absorbed through the walls of the hyphal cells. The movement of molecules through the cell wall and cell membrane is done by vesicular transport—dissolved nutrients and enzymes are carried between the cell and the space outside it in a membrane-bound pocket or vesicle. This is an active rather than passive process, consuming energy, as the cells need to strictly regulate the fluids that enter and leave them—both dehydration and overhydration would be harmful. Parasitic fungi attack and feed on living cells in a similar way, both plant or animal (and even other fungi in some cases). They invade their host by secreting digestive enzymes, which break through the host's cell wall (in the case of plants and fungi) and cell membrane and allow the growing fungal hyphae to enter the cell.

The organelles of fungal cells are generally similar to those of animals—they lack plastids, but possess mitochondria for energy generation and other functions, including regulating cell maturation and cell death. They use ribosomes to assemble proteins from amino acids, the Golgi apparatus that helps prepare enzymes and other proteins (including cell wall constituents) prior to transportation through the cell membrane, and their nucleus contains the DNA instructions for protein coding.

Below *Bracket fungi will grow and feed on on living trees, and may eventually cause enough damage to kill their host.*

Below *Structure of fungal cells at the tip of a hypha.*

Fungal cell

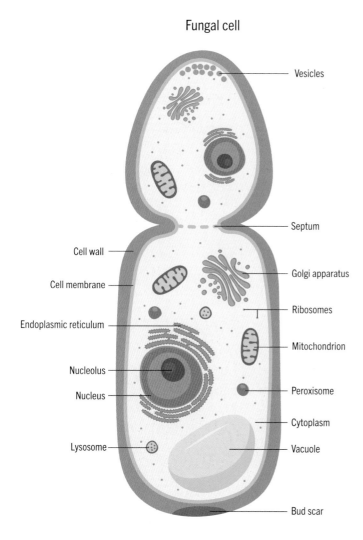

Vesicles

Septum

Cell wall

Cell membrane

Golgi apparatus

Ribosomes

Endoplasmic reticulum

Mitochondrion

Nucleolus

Nucleus

Peroxisome

Cytoplasm

Lysosome

Vacuole

Bud scar

PREDATORY FUNGI

While the majority of fungi absorb nutrients from decaying organic material that is in or on the soil, the group includes those that obtain food via a symbiotic link with another species, or through parasitism. More surprisingly, the group also includes a few that capture and consume living animal prey. Predatory fungi are much less conspicuous than predatory plants, as their activity is underground or within leaf litter, but their methods are no less ingenious. Some grow tiny ring-shaped structures on their hyphae, which capture tiny, thread-like nematode worms by constricting around their bodies when they try to move through the ring. Others produce sticky pegs on their hyphae, to snare passing springtails, rotifers, and other very small animals. Once a prey item is caught, the hyphal cells secrete enzymes to break it down and consume it in the usual external way.

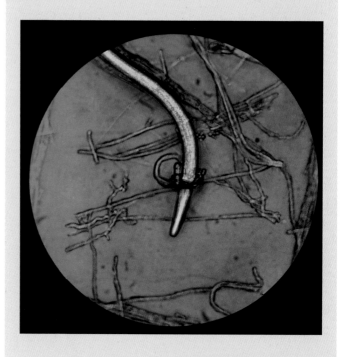

Above *A predatory fungus snares a nematode worm.*

// Reproduction in fungi

Mushrooms and toadstools can be as colorful and striking as flowers. When animals like us notice and disturb these fruiting bodies, perhaps by picking and eating them, we can help them to disperse some of their extremely numerous spores, in a similar way to how animals can help disperse plant seeds by shaking stems and eating fruits. However, fungi do not depend on the help from living animals to fertilize them.

An individual fungus can be incredibly effective at spreading itself over a wide area. Hyphae can theoretically grow forever, so habitat changes and resource shortages may be the only limitations they face. Generating new individuals is, though, a more effective way of propagating your genes across a wider area. Fungi can also propagate via fragmentation—if a piece of soil is moved from one place to another, any mycelia it contains can continue to live and grow in the new area. When it comes to "intentional" reproduction, many fungi do this sexually as well as asexually—both methods result in the creation of spores, distributed from sporangia. Each spore holds the potential to grow into a new individual organism.

With asexual spore generation, the spores develop in an individual fungus's sporangia (which grow either on hyphae or within fruiting bodies) without any genetic input from another individual. Sexual reproduction occurs when cells of two different mating types (analogous to the two sexes we typically see in animals and plants, but occurring in more than two variants) come together. This may occur through the production of gametes, which form inside structures on the hyphae called gametangia, but other fungi dispense with gametes and the actual hyphae simply fuse. They can then combine genetic material and create spores that are a genetic combination of both parents.

Individual spores, regardless of the size of the fungus that produces them, are extremely small, averaging 2–50 micrometers across. One *Agaricus* mushroom can produce several billion spores, although only a tiny proportion of them will germinate successfully under typical conditions. Each is a single cell with a sturdy coat, containing all the organelles and other material it needs to grow a hypha and begin to acquire food for itself.

Below *Each fruiting body of the fly agaric fungus produces billions of white spores, each just 9–13 by 6.5–9 micrometers.*

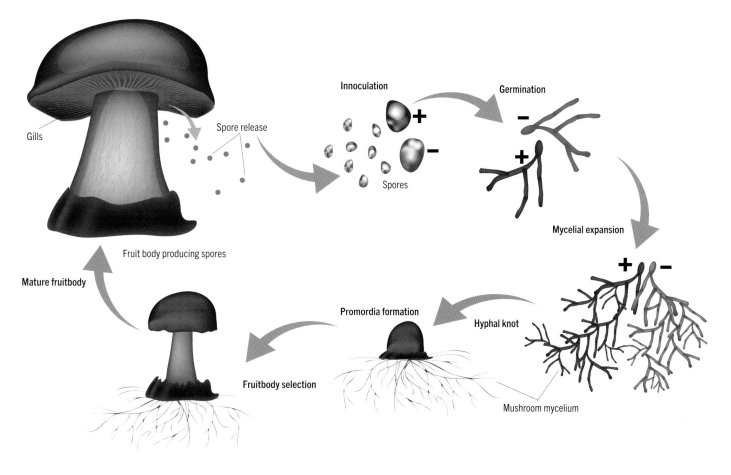

Gills

Spore release

Innoculation

+

−

Spores

Germination

−

+

Mycelial expansion

Fruit body producing spores

Mature fruitbody

+ −

Fruitbody selection

Promordia formation

Hyphal knot

Mushroom mycelium

Above *The life cycle of a mushroom.*

THE HUMUNGOUS FUNGUS

As the only one of the three familiar kingdoms of life to have unicellular and multicellular forms, the fungi truly do embody the idea of all creatures (or all organisms, to be precise) great and small. We have met the smallest, the single-celled yeasts (see page 96). Those who want to meet the biggest need to travel to Oregon, USA, where a single individual of the honey fungus *Armillaria ostoyae* has grown, over as much as 8,000 years, across an area of more than 2.9 miles^2 (7.5 km^2). The combined weight of its innumerable, microscopically slender hyphae is estimated at somewhere between 8,270 and 38,580 US tons (7,500 and 35,000 tonnes), making it easily the largest individual organism on the planet. This species is a tree pathogen and over its long life will have killed thousands of trees, generating a vast amount of soil nutrients for itself and many other organisms.

Above *Honey fungus can be a very destructive species and is unwelcome in arboretums and commercial plantations.*

// Fungi in ecosystems

When we consider a food chain or food web within a thriving ecosystem, we look at the sunlight-harnessing, soil-sucking plants, the herbivorous animals that eat them, and the carnivorous animals that eat the herbivores. Then it is back to the beginning with plants taking up nutrients released from the decaying remains of dead plants, dead animals, and animal droppings. What we tend to neglect are the organisms that complete the cycle, by breaking down all of that dead and waste material and returning nutrients into the ground in a form that the plant roots can use.

This group of organisms are called saprotrophs or saprophytes (from *sapro*—Ancient Greek for rotting matter), or sometimes detritivores. They are represented by a huge variety of species from across all domains of life, including vast numbers of prokaryotic organisms. Some of these have the vital role of last-stage breakdown of nitrogen-containing compounds to the simple molecules that plants require, while larger eukaryotes such as earthworms are active at the other end of the process, pulling fallen leaves down into their below-soil burrows. Earthworms feed by swallowing soil and then excreting it—their castings are very rich in nitrogen-containing compounds, but also improve the soil aeration and structure for plant roots and add numerous beneficial bacteria to the soil.

Soil-dwelling organisms may be part-time predators as well as saprotrophs, and of course they also produce their own waste materials and leave behind their own carcasses, adding to the nutrient richness of the soil they live on or in. Collectively, they form their own food web, much of it operating on a microscopic scale. Fungi are a key part of this community and are hugely important by virtue of their great abundance as well as their mycorrhizal links with plants and other associations with a range of species.

Fungi that parasitize other organisms can also be of great ecological importance, and we make use of some species that attack insects and nematodes in biological pest control. The genus *Cordyceps* is especially well-known as an insect pathogen—species of this group invade and eventually kill their insect hosts, but may first manipulate the host's behavior, causing it to climb to a high point before it dies, for effective dispersal of the spores released from the tiny fruiting body that erupts out of the host. Many plant diseases are caused by fungi, but are themselves attached by other fungus species—these species play an important natural role in helping plants to resist infections and they too can be used in biological pest control. Several species of the genus *Trichoderma* have proved effective controllers of *Botrytis* (grey mold) infections and various fungal root-rot diseases.

Above left *Insects killed by* Cordyceps *infection are often found clinging to plants in exposed places, with delicate fruiting bodies growing out of them.*

Below left Trichoderma *is a soil-dwelling fungus which attacks a range of other fungal species, and is widely used as a biological pest control agent.*

NOBLE ROT

Few commercial growers want to see their plants attacked by fungus. However, infection by the "noble rot" mold, *Botyrtis fuckeliana*, is a vital part of cultivating the grapes used to produce sweet white wines such as Sauternes. Only when the grapes are covered in the alarming-looking fluffy white fungal growth are they ready to be harvested and turned into wine. The fungal hyphae invade and dehydrate the growing grapes, resulting in a higher concentration of the flavorsome sugar and acid molecules that give the wine its distinctive powerful flavor.

Above *Grapes affected with "noble rot."*

Above *A mycorrhizal network.*

// Microbial symbiosis

The concept of symbiosis has cropped up repeatedly in these pages, from the first endosymbiotic events that kicked off the evolution of eukaryotes, to the ubiquitous but astonishing subterranean network of co-operation that exists between higher plants and fungi. Complex life on this planet arose out of community, and so it continues to this day. Associations between different species aren't always harmonious—predation, parasitism, and competition all work to the detriment of some individuals, but all are important driving forces of evolution and natural selection, too, and the same goes for mutually beneficial relationships, whether on the macroscopic or microscopic scale.

Just as adapted cyanobacteria now provide photosynthesis for green algae, so green algae provide the same service for the occasional animal. The species *Elysia chlorotica*, an alga-eating sea slug, retains a living population of chloroplasts in its gut, harvested from the algal cells. The slug benefits from the chloroplasts' ongoing photosynthesis, although it's fair to say that the alga itself is not a winner in this arrangement. There is only one known example where an alga actually exists as an intact, living endosymbiont inside an animal's cells—the host is the spotted salamander in its egg stage. *Chlorococcum* algal cells live in the salamander's cells as they divide, their photosynthesis giving oxygen that boosts the embryonic amphibian's development, while the algal cells consume the carbon dioxide that the salamander embryo releases.

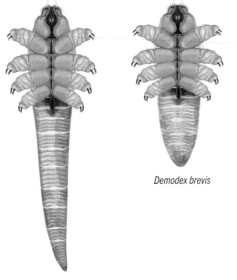

Demodex brevis

Demodex folliculorum

Above *Two species of* Demodex *mites, which live on human skin.*

Above *The green coloration of the sea slug* Elysia chlorotica *is derived from chloroplasts from the algae it eats, and these chloroplasts continue to function in their new host.*

Our reliance on the microbe inhabitants of our guts is increasingly well understood, but we and other mammals also benefit from the community of microorganisms that live on our skin. Among the most unappealing of these organisms are the eyelash mites, *Demodex*, microscopic relatives of spiders that live in and around our eyelash hair follicles and skin sebaceous glands. They may cause irritation and inflammation when too numerous, but on the whole are thought to be useful, cleaning up dead cells and excess sebum (skin oil) and thus preventing more damaging microbes from setting up home on and in our skin.

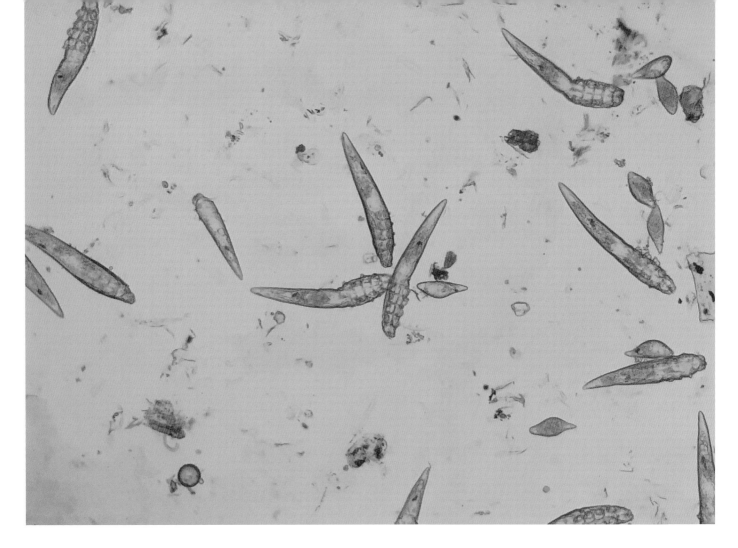

Above Demodex folliculorum *under the microscope.*

Above *Spawn of the spotted salamander, colored green by its symbiotic algae.*

TIPPING THE SCALE

Recent studies have shown that the algae that live within the embryonic cells of spotted salamanders do not get as much out of the arrangement as their host does, as they struggle to meet their own oxygen needs when in endosymbiont mode. This is not unexpected. It is unlikely that any relationship of this nature between two species is perfectly equitable, because natural selection favors those individuals that do the best for themselves, regardless of whether this is to the benefit or the detriment of any symbiont they may happen to be involved with, so over time many mutualistic relationships slide more towards parasitism of one on the other. Even within the same species pairs, it's often possible to find some individuals enjoying equal-ish benefits from the partner species, and others very much not. In evolutionary terms, a species successfully behaving more parasitically towards its host is under selective pressure to continue the association, while the one being exploited is under selective pressure to change or escape it.

// Microscopic animals

If you have ever taken a look at a drop of pond water under a microscope, you have probably been fascinated by the likes of water fleas and copepods—the charming diminutive cousins of familiar crustaceans such as crabs and lobsters. If you choose to learn more about *Demodex* mites, introduced on page 166, you may well wish that you hadn't, because it is quite disturbing to consider that actual animals (and rather alarming-looking ones at that) live inside our hair follicles, and wander across our faces by night. The fact is that our world swarms with miniature animal life of many kinds, all of them only known to us because of microscopy.

Some phyla of animals are entirely microscopic. Perhaps the best known of these are the tardigrades. Most species are no longer than half a millimeter when fully grown, but a low-powered microscope reveals the detail of their segmented, eight-legged bodies. These squat and slow-moving animals feed mainly on bacteria and plant matter. They are active in water films on mosses and lichens but can live for long periods in a dehydrated state. In this state, they are noted for their extreme resilience to the sorts of conditions that would kill off nearly all other animals—extremes of temperature, powerful toxins, pressure, and radiation. Dehydrated tardigrades were taken into a low Earth orbit on an astrobiology space mission in 2007 and exposed to the vacuum and solar radiation of outer space. More than half of them survived the experience and were even able to go on to reproduce.

Another phylum of microscopic animals is Rotifera. These water-dwelling animals are so called for their "crown" of cilia at the head end, which beat rapidly in a manner suggestive of a turning wheel as they sweep water (and food particles) into the mouth. Some rotifers can swim and others are immobile (sessile). Rotifers generally have a set number of cells in their bodies when adult—about 1,000 in most species. Despite this very limited cell count, their bodily systems include a brain and nervous system (including up to five eyes), a digestive tract with several distinct sections, and male or female reproductive anatomy.

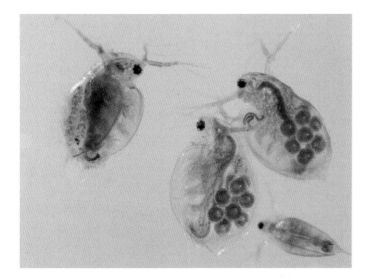

Above *Familiar to fish-keepers as a food source for their pets, these* Daphnia *are minuscule crustaceans.*

Right *The largest animal on Earth, the blue whale, feeds on some of the smallest, eating about 16 tons of plankton each day.*

Many nematode worms are microscopic throughout their lives. These incredibly abundant animals are parasitic in some cases but there are also many free-living species, which are present in virtually every habitat. Most members of the obscure phylum Loricifera are also microscopic. These are sessile animals that live on gravelly seabeds, extending feeding appendages from a tough outer casing.

Arthropoda (a huge phylum of "jointed-legged" animals, including insects, spiders, centipedes, and crabs) includes a number of microscopic species such as the pond-water crustaceans mentioned above, as well as *Demodex* and many other mites, which are related to spiders and scorpions. Many other arthropods that are small as adults are microscopic in their egg and early larval forms—this includes insects and many aquatic arthropods as well. Marine zooplankton comprises large numbers of these, as well as the microscopic larvae of fish, tunicates, jellyfish, molluscs, and more.

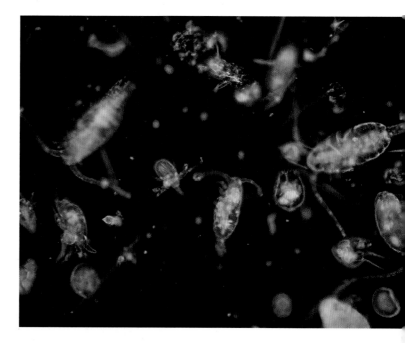

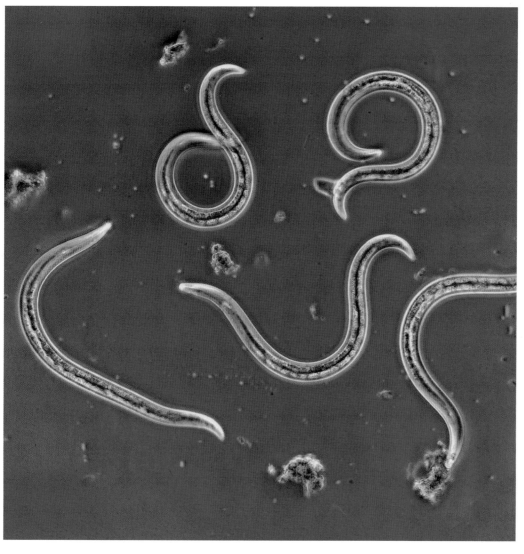

Above *Copepods, tiny crustaceans which are a key component of marine plankton.*

Left *Nematode worms are mostly microscopic and hugely abundant, with as many as 10 million individuals per cubic meter of soil.*

// Zooplankton

The innumerable tiny living things that swim or drift in open water are known as plankton, which is no more precise a term than "water life," but what unites them is that they have limited swimming power. Marine plankton, therefore, are rather at the mercy of ocean currents. They are traditionally divided into two groups—the zooplankton (animal-like) and phytoplankton (plant-like). The phytoplankton are those that photosynthesize, but the grouping still comprises a lot of taxonomically unrelated lineages, including diatoms, dinoflagellates, and various algae. The zooplankton include unicellular protozoa, multicellular but microscopic animals, alongside the very early life stages of other animals that will, when they are adult, be far from microscopic.

The plankton form a dynamic community, in both fresh water and the seas. Their daily cycle is governed by the movement of the phytoplankton, which move close to the surface in daylight in order to capture more sunlight. Zooplankton feed on phytoplankton (and each other, often including others of their own species) so they follow the phytoplankton. The exception to this feeding chain is fish eggs, which carry their own supply of yolk and so do not need to eat anything else—this yolk will sustain them in the days after hatching, as well.

Above *This 0.3 in (7.5 mm) tilapia larva is about to emerge from its egg. It carries a food supply in the form of a bulging yolk sac.*

Below *Plankton are tiny, but their great abundance means they can support some of the sea's largest animals, including the biggest fish of all—the whale shark.*

Above *A crab zoea, bearing little resemblance to its eventual adult form.*

The larval constituents of zooplankton are often drastically different in appearance to their older selves. Newly hatched fish larvae look odd and disproportionate, with their big heads and obvious bulging yolk sacs, but underdeveloped fins and skinny hind bodies. The larva of a crab, though, looks entirely different and, unsurprisingly, these larvae were classed as different species and genera to adult crabs when first observed. This stage, called a zoea, has a long segmented hind body with which it swims, large eyes, prominent gills, and two long spines on its small carapace: a central one pointing up and another pointing down. As it matures and passes through a series of molts, the crab's carapace broadens, eventually covering up its abdomen, and it becomes a seabed walker rather than an ocean drifter.

WHY SO SMALL?

The bluefin tuna is a popular food fish for humans. If you are used to seeing it in tins, you may be startled by the size of the adult fish, which after ten years of growth can weigh more than 0.5 US tons (500 kg) and reach 9.8 ft (3 m) in length. Even more startling is that this monster grew from a near-microscopic, 0.1 in (3 mm)-long hatchling larva. Needless to say, only a tiny proportion of these larvae will reach full size, and only a tiny proportion of the millions of eggs produced by a spawning female will even make it to larval stage. Animals that produce vast numbers of tiny offspring are known as "r-selectionists." Their investment per offspring, in time and bodily resources, is minuscule, and the degree of wastage is colossal. Humans and many other mammals, by contrast, are "K-selectionists." We usually have one offspring at a time, and invest months nurturing it *in utero*, followed by years of intensive parental care as it slowly matures. These two very different strategies often actually have similar outcomes, in terms of number of young per parent that survive until breeding age.

Above *Bluefin tuna produce vast numbers of microscopic eggs, of which only a minuscule proportion will survive to adulthood.*

// Animal tissue types

The Visible Human Project, which you can view via this link: www.nlm.nih.gov/research/visible/visible_human, presents a huge collection of photos of the human body. Specifically, these photos show 0.04in (1mm) cross-section slices through one male and one female human cadaver. These images naturally trigger a bit of squeamishness, but if you can overcome that, they provide an extremely interesting insight into our bodies, especially the appearance and distribution of our various tissue types.

For all our many organs and their various constituent parts, we humans, along with other complex animals, have only four basic tissue types (but many modifications of these). The four types are: epithelial; connective; muscle; and nervous. All types take somewhat different forms and arrangements in different animal groups—for example, as well as a distinct brain, insects also have a smaller brain-like bundle of nervous tissue (a ganglion) in each of their body segments, which handles certain activity in that segment alone. We can easily point to where muscle and nervous tissue can be found in our and most other animals' bodies, but epithelial and connective tissue are more diverse and obscure, and can be trickier to define.

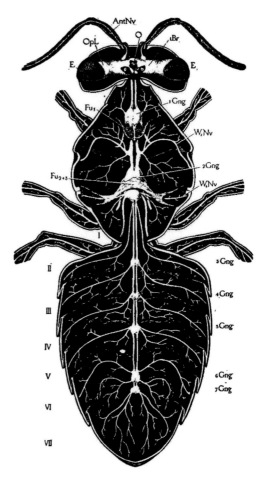

Above right *Diagram of the nervous system of a typical insect, comprising a series of ganglia (Gng) and nerves (Nv).*

Below *The early stages of cell differentiation of a human embryo, soon after fertilization.*

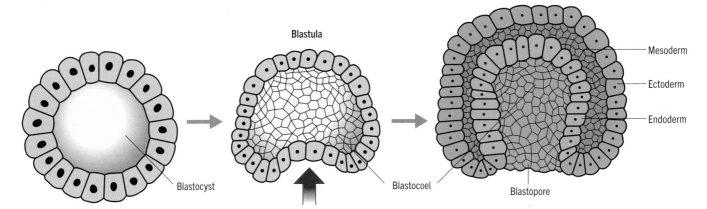

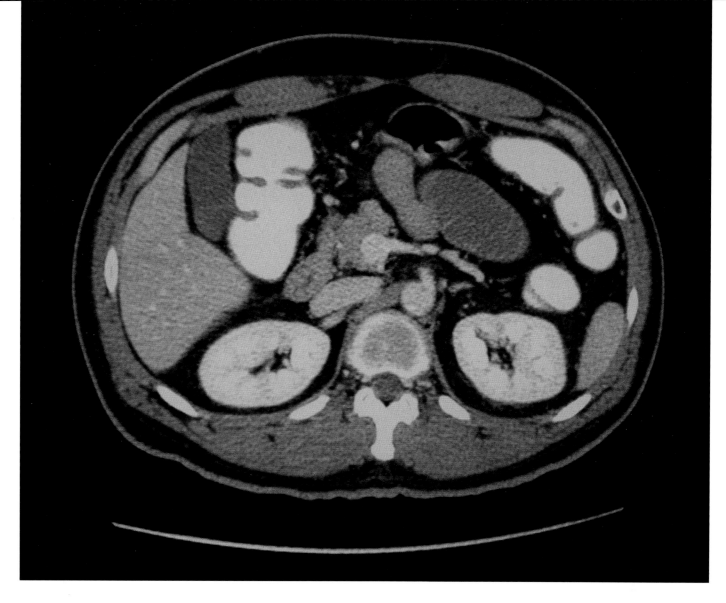

Above *A CT scan image of a cross-section through a human abdomen, showing several organs and tissues.*

Epithelial tissues form protective linings around organs and the vessels in vascular systems—depending on location, they may also secrete enzymes and hormones. Most of our internal organs are formed primarily from epithelial tissue (a notable exception is the heart, which is composed of a specialized type of muscle tissue). The connective tissues in our bodies provide structure and conduits for distribution of nutrients—they include bone, adipose tissue (fat), and our blood. In invertebrates, connective tissues form the exoskeleton or other parts that are involved with supporting the body (but note that the shells of molluscs are not living or formerly living tissues—they are secreted mineral material). All of these four types are differentiated into a range of variants—for example, our bodies contain skeletal, smooth, and cardiac muscle, all of which show distinct cellular shapes and arrangement under the microscope, but the cells themselves share a set of particular traits that make them all recognizably muscle cells.

Cells that are yet to differentiate into particular types and form tissues are known as stem cells and those that are in the early stages of becoming differentiated (but already committed to a particular tissue type pathway) are called progenitor cells. In an early-stage vertebrate embryo, the first differentiation of stem cells is into three germ layers— the ectoderm, which is destined to form certain outer epithelial tissue and also nervous tissue; the mesoderm, from which deeper epithelia and muscular and connective tissues develop; and the endoderm, which forms the epithelial tissue that makes up many internal organs. Stem cells continue to be produced in certain parts of the body throughout life, as the cells that comprise some tissue types have short lifespans and are replaced numerous times over the organism's lifetime.

// Animal respiration

The word "respiration" is often used in common parlance to mean breathing. The device known as a respirator is a complex mask, used in dangerous settings to prevent the wearer from inhaling hazardous gases and particles. In biology, respiration refers to the chemical reaction that occurs in organisms' cells to create chemical energy (ATP molecules) through the breakdown of stored (or newly generated) glucose molecules. As we have seen, all living things respirate, and in most animal respiration oxygen is consumed and carbon dioxide is produced as a by-product. The movement of these two gases is central to keeping the respiration process going, and so biological respiration is indeed associated with breathing in and out.

The site of respiration in animal cells, as in other eukaryote cells, is the mitochondrion. Cells that require a lot of energy in their day-to-day functioning, such as those that form muscle tissue, have a rich population of mitochondria. In human cardiac muscle cells, which are constantly active in the beating heart, as much as 35 percent of the interior cell space may be taken up with mitochondria, while for most skeletal muscle this falls between 8–12 percent. Red blood cells, by contrast, have no mitochondria, although their progenitor cells do. These cells do not, therefore, consume oxygen through aerobic respiration—instead, their role is to carry it to where it is needed. They do have an anaerobic biochemical pathway for energy generation, though.

In our bodies, these gases are transported in the blood, and move into and out of the bloodstream via the alveoli in our lungs. We have about 480 million of these microscopic structures in our lungs. Each alveolus is about 200 micrometers across, and is surrounded by blood capillaries with walls just one cell thick, through which oxygen and carbon dioxide can diffuse. Incoming oxygen molecules attach to red blood cells, while carbon dioxide travels the other way. The diffusion works in this direction because of the differing partial pressures of the two gases in the blood and the alveoli—gas molecules naturally move from areas of high concentration to areas of lower concentration.

Opposite *When we push ourselves very hard, our bodies may switch to anaerobic respiration, but this method of energy release is not sustainable for very long.*

Below *The process of gas exchange in lung tissue.*

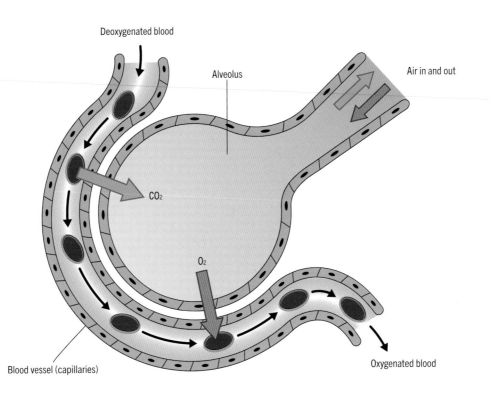

Deoxygenated blood

Alveolus

Air in and out

CO_2

O_2

Blood vessel (capillaries)

Oxygenated blood

ANAEROBIC ALTERNATIVE

Every active cell and every biochemical process has its working limits, beyond which efficiency will break down and things may stop altogether. You will know that when you are really exerting yourself, your heart begins to pump very fast, and your breathing rate also increases. Both of these changes occur because your muscle cells are carrying out respiration at a high rate, and need to have extra oxygen delivered to them through your blood. If this process can't keep up, you will be unable to continue with the exertion, but your cells do have a last-ditch alternative biochemical trick to give you a last push of energy (which may be just enough for you to complete your workout, or escape from the tiger that is chasing you). Part of the glucose breakdown process can be carried out without oxygen. This anaerobic activity generates the by-product lactate and also releases hydrogen ions, which causes the acidity of your blood to increase quickly and create a burning sensation in your muscles. This form of respiration can only be sustained for a couple of minutes before your body forces you to stop.

// Animal nervous system

When we see a tasty snack, we reach for it, eat it, and enjoy the flavor. If we see a dangerous predator bearing down on us, we run away (or climb, or hide) to save our skin. Finding food while avoiding becoming food is what keeps animals alive day on day, and to do these things they need a nervous system. This enables them to perceive the world around them, make decisions on the basis of that information, and physically move their bodies with intentionality. We humans consider that our big brains and high intelligence set us apart in the animal kingdom, but the nervous system is more universal than we might think. Many very simple microscopic animals, which lack any equivalent to the other organs and systems we have, have devoted a sizeable proportion of their handful of body cells to nervous tissue, so great is its importance to survival.

The nerve cell or neuron is one of the most distinctive types of cells in our bodies. It has three obvious parts—the cell body or soma, where the nucleus and other organelles reside and which bears numerous branched projections called dendrites; the axon, which is a single very long, fine thread projecting from it; and the axon terminal, a series of branched, bulb-tipped fibres at the axon's far end. The purpose of this cell is communication, and in our bodies a signal may pass through many nerve cells in between its origin and its destination. However, some individual human neurons have extremely long axons—those that are bundled together to make up our sciatic nerve (running from hip to heel) can be 3.3 ft (1 m) long, although are only about 0.5 micrometers wide.

Below *Structure of a neuron.*

Right *The human "knee jerk" reflex involves sensory neurons receiving an input, passing this information to the brain, and the brain sending out its response through motor neurons which trigger a rapid muscular contraction. The process happens at high speed and with no involvement from the conscious brain.*

A nerve impulse or action potential is an electrical impulse, which travels through a neuron's cell membrane, making its way to the axon terminal of the cell via the axon. From there, the signal crosses the space (synapse) between the terminal bulbs and the dendrites of the next neuron. This crossing can be chemical, whereby the bulbs release molecules called neurotransmitters, which bond to receptors on the next neuron's dendrites, or it can be electrical, with a flow of charged sodium and potassium ions causing a voltage change in the bulbs, which is passed on to the next neuron. Conduction of a nerve impulse along an axon is very fast, and further accelerated by the presence of a myelin sheath. This fatty covering, secreted by specialized Schwann cells, both protects and insulates the axon, with the electrical charge "jumping" along via tiny gaps (nodes of Ranvier) between the sections of myelin.

Neurons communicate with other cell and tissue types constantly. In the case of humans and other mammals, neurons receive input from sensory receptor cells in our ears, eyes, mouth, nose, and skin, and pass these signals to the brain. We may consider this data consciously, or react to it unconsciously. Outbound nerve impulses stimulate our skeletal muscle cells to contract, which can cause a joint to bend or straighten, trigger epithelial cells to secrete hormones, and regulate the pace of our heartbeat, among many other bodily responses.

Above *Sensing one's surroundings and the state of one's internal environment is key to survival. Even sponges, perhaps the simplest of all animals, have a rudimentary nervous system.*

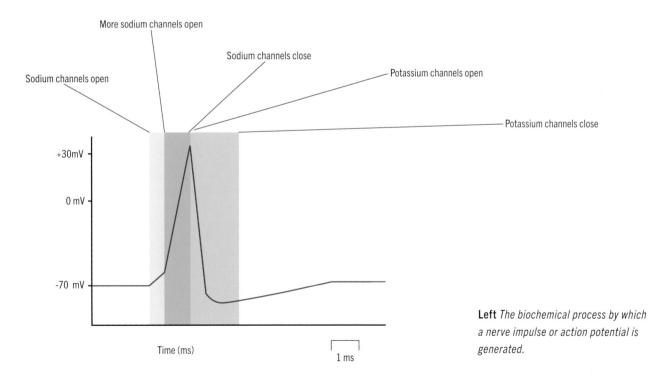

More sodium channels open

Sodium channels close

Sodium channels open

Potassium channels open

Potassium channels close

+30mV

0 mV

-70 mV

Time (ms)

1 ms

Left *The biochemical process by which a nerve impulse or action potential is generated.*

// Animal connective tissues

The earth has a mass of 6,613,867,865,546,327,000,000 US tons (6,000,000,000,000,000,000,000,000 kg) which means coping with a gravitational force of 32.2 ft/s^2 (9.81 m/s^2). This means that smaller objects within Earth's gravitational field are pulled towards the center of the planet hard and fast. Gravity prevents us from drifting off into space, but it also requires us to have bodies that are sturdy, strong, and held-together enough to stand up and move about, in defiance of that gravitational pull. Connective tissues provide us with this strength. The nature of our connective tissues also places limitation on our body size, and because buoyancy in water provides support of its own, water-dwelling animals have the potential to grow to larger sizes without a need for such strong connective tissues.

Hard and dense tissues provide the necessary strength and rigidity. Vertebrates like us possess an internal skeleton made of bone, which contains cells called osteoblasts. These cells secrete minerals (primarily calcium) around a collagen matrix, forming bone tissue. Once they are surrounded by this material, the osteoblasts become osteocytes, which are held in place in the matrix and no longer produce bone but have various metabolic functions around maintaining bone tissue (including helping to heal any fractures that occur). In cross-section, a typical bone in the human body has dense and compact bone tissue around the outside, with looser, more metabolically active spongy tissue, including bone marrow, on the inside. The cells in our blood (another type of connective tissue) are mainly synthesized in bone marrow.

Below *Different types of bone cells.*

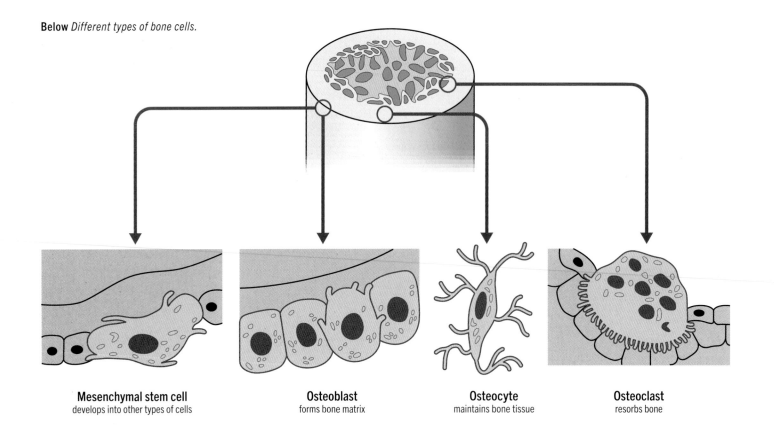

Mesenchymal stem cell
develops into other types of cells

Osteoblast
forms bone matrix

Osteocyte
maintains bone tissue

Osteoclast
resorbs bone

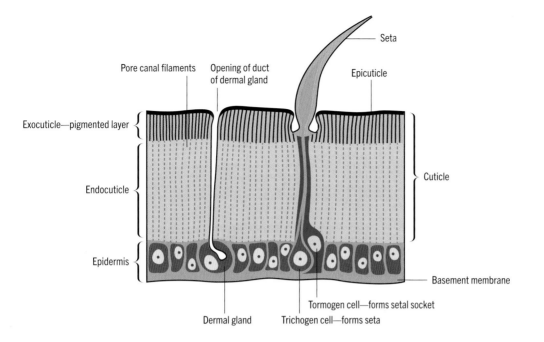

Pore canal filaments
Opening of duct of dermal gland
Seta
Epicuticle
Exocuticle—pigmented layer
Endocuticle
Cuticle
Epidermis
Basement membrane
Tormogen cell—forms setal socket
Dermal gland
Trichogen cell—forms seta

Left *Structure of an insect cuticle.*

Below *A copepod, minuscule water-dwelling representative of the phylum Arthropoda. This female individual will carry her two sacs of eggs with her until they hatch.*

Arthropod invertebrates, which include insects and other important groups of land-dwelling animals, keep their supportive connective tissues on the outsides of their bodies, in the form of an exoskeleton. The outer layer (cuticle) of this is non-living material made mainly from the protein chitin, interspersed with mineral molecules, all secreted by epithelial cells beneath. This jointed exoskeleton is important to prevent water loss from the body, but also gives insects enough body strength to walk, hop, climb, run, and fly. It is not made of living cells though, so cannot grow. Arthropod growth, therefore, requires the animal to molt its entire exoskeleton, emerging in a new and temporarily softer cuticle, which expands to accommodate the bigger body before it becomes rigid.

SCALING UP AND DOWN

Land-dwelling arthropods are never very big compared to larger vertebrates. Part of the reason for this is the nature of their exoskeletons. This material provides rigidity but its density makes it considerably heavier than an equivalent volume of bone would be. Small muscles have a more effective power-to-weight ratio than larger ones, and if an insect were the size of a human, its muscles could never be strong enough for it to move its own mass around. Although relatively lightweight, bone is still denser than other tissues. Birds, which have many adaptations to reduce their mass relative to body size, have a minimalized skeleton relative to similar-sized mammals, with fewer and smaller bones, to help them win their battle against gravity.

// Animal reproductive systems

Most animals reproduce sexually, bringing male and female gametes (haploid sex cells) together to form a new cell, a diploid zygote. This one cell has the potential to divide, differentiate, and mature into a new individual animal, carrying a combination of its parents' genes. We have already seen how this process can work in various other organisms, both unicellular and multicellular. In the case of animals, many have evolved specialized tissues and structures that modify the process of fertilization, and allow one or both parents to provide extended care and protection to the zygote as it develops.

In a few cases (for example, snails and earthworms), the same individual animal can produce both male and female gametes, and in a few others (certain fish and crustaceans) they may change the type of gametes they make during their lifetime—these are known respectively as hermaphrodites and sequential hermaphrodites. However, more often an animal is of one sex, which is determined by the chromosomes present in the gametes that formed it and is fixed for life. As it matures, it develops the reproductive anatomy to make one gamete type only. Female animals make female gametes or ova, which are large, complex, and non-motile, and males produce the much smaller, motile male gametes or sperm.

Below *Male frogs release sperm into the water as the female releases her unfertilized spawn. A male improves his chances of being the one whose sperm finds its target by locating a female and holding tightly onto her until the process is over.*

Above *Technically, each ostrich egg is a colossal single cell.*

In most animals, gametes develop from progenitor cells that form inside structures called gonads—in males these are testes and in females they are ovaries. The final round of cell division during gametogenesis is meiosis rather than mitosis, so the final products have a half-set of chromosomes. Then it is a matter of bringing sperm and ovum together. Many water-dwelling animals use external fertilization—both sexes release their gametes into the water, and sperm cells seek out ova (the former usually outnumber the latter by a sizeable degree so all ova are likely to be fertilized). Sperm cannot swim on dry land, and both gametes and zygotes would desiccate, so land animals use internal fertilization. Males use a penis or phallus to release sperm inside the female's reproductive tract. Very rapidly after fertilization, the new zygote begins to divide and soon becomes a ball of cells that start to differentiate into fundamental tissue types, destined to form epithelia, connective tissue, muscle, and nerve tissue of various kinds.

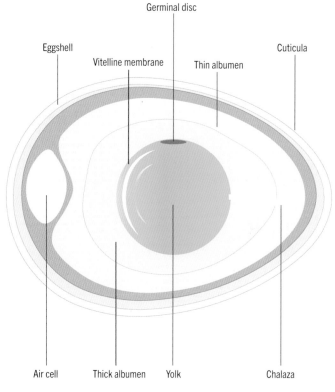

Above *Structure of a chicken egg. The germinal disk occupies just 1 percent of the volume of a newly laid chicken's egg, and contains the nucleus and all other cell organelles.*

Most land animals lay eggs, with the embryo enclosed in a much larger protective case or shell of tough protein or mineral-based material to prevent dehydration. Insect eggs, though tiny, can be spectacularly beautiful when viewed under magnification, with elaborately textured shells that have evolved to provide camouflage or discourage predators. Birds and reptiles lay large, shelled eggs, filled with non-cellular food and water supplies (yolk and albumen respectively—technically, a 3 lb (1.4 kg) ostrich egg is a single cell). The embryo is connected to this, and can therefore advance its development considerably before hatching.

A female placental mammal retains each fertilized ovum in her uterus, and connected to her blood supply via a placenta, which deals with feeding the growing embryo and removing its waste. She gives birth when the embryo's development is sufficient that it can survive outside of her body—this state of development, and the nature of parental care after birth, varies hugely between different mammal species.

Above *The pinhead-sized egg of a brown hairstreak butterfly: magnification reveals its elaborate structure.*

// Animal metabolism

Animals have the same needs for carbon, nitrogen, oxygen, and hydrogen, along with trace amounts of other elements, as other organisms. All animals are consumers or heterotrophs—they must eat organic material to obtain their nutrients. As we have seen, the very simplest of sea animals take food particles directly into the cells on the outsides of their bodies through phagocytosis, but nearly all animals do have some form of digestive tract, with a mouth at one end, an anus at the other, and a lining of cells in between that produce digestive enzymes and absorb nutrients. In more complex animals, the digestive tract contains multiple organs and cell types, and often specialized structures or regions that enable them to consume a similarly specialized diet.

Mouths are the entry point for food, but can also be an initial site of food breakdown. For us, this means biting and chewing with our teeth, made of mineralized, bone-like tissue, which is the hardest material in our bodies, and releasing the enzyme amylase, which gets to work on carbohydrates in the food. Molluscs have a ribbon-like structure called a radula, covered with tiny "teeth" of hard chitin, which they use to rasp layers of cells away from what they are eating (for example, algae growing on a rock). Insects have multiple mouth parts formed from jointed appendages, which are adapted to work in a variety of ways—in butterflies, they form a sucking tube; in flies, a sharp stylus that pierces as well as sucks, and beetles have sideways-mounted jaws that are used to grab and bite.

Below *Absorption of food molecules through a villus, thousands of which line the human small intestine.*

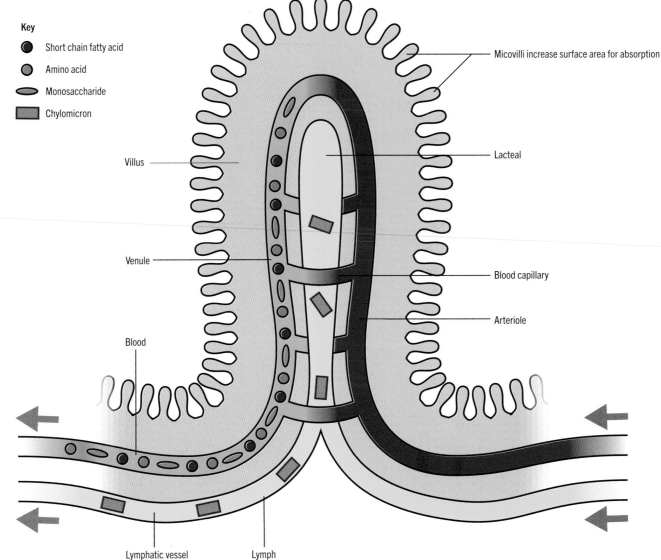

Key
- Short chain fatty acid
- Amino acid
- Monosaccharide
- Chylomicron

Micovilli increase surface area for absorption

Villus

Lacteal

Venule

Blood capillary

Arteriole

Blood

Lymphatic vessel   Lymph

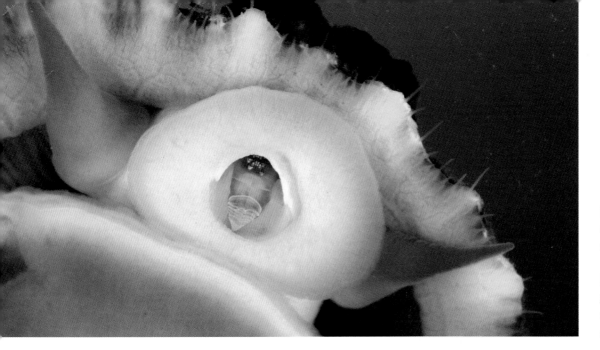

Left *The rasping mouth or radula of a limpet, designed to scrape algae from the surface of rocks.*

Below *Microscope view of glomeruli, the structures through which blood is filtered in the mammalian kidney.*

Within our stomach and gut, various enzymes (including proteases, lipases, and amylases, attacking proteins, fats, and carbohydrates respectively) carry out further breakdown. Some digestive enzymes are secreted by cells in the gut itself, while the pancreas produces others. Once small enough, food molecules are absorbed by the epithelial cells that cover the tiny finger-like projections (villi) lining the sides of our small intestine. They are then further processed within the cells or elsewhere—amino acids may be used to build proteins that the cell requires, while excess glucose molecules travel to skeletal muscle and the liver, where they are built into a storage polymer called glycogen.

Maintaining the correct water balance is key to the survival of individual cells and the organism as a whole. Vertebrates have kidneys, containing minuscule tubular structures called nephrons. Fluid can move between nephrons and blood capillaries, and the nephrons serve as filters for the blood, extracting excess water and certain waste materials.

NOT SUCH A WASTE?

Not everything that enters the mouth can be used by the body, even in animals with highly efficient digestive systems. Animals therefore produce waste, but what is unusable by them can still be a useful nutrient source for other species. Examining a sample from a cow pat under the microscope reveals a thriving population of prokaryotes—in a fresh cow pat, these are mainly the cellulose-fermenting species you find in the cow's digestive tract, but over time they are replaced by a new community of saprophytic species, able to utilize what the cow and its microbiome could not. Many insects lay their eggs on animal droppings too.

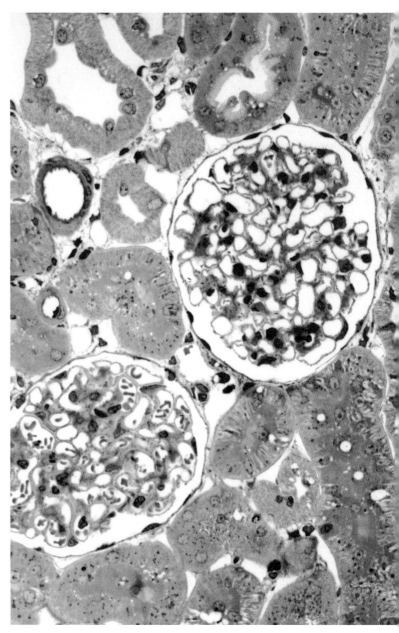

// Animal movement

There are few more inspiring sights in nature than animals on the move, whether they be galloping horses, swirling starlings, a gracefully pulsing jellyfish, or a fast-flashing shoal of fish trying to dodge a speedy shark. The generation of these various movements all involve interplay between different cell and tissue types, under the control of the nervous system, and often require cells to show extremely rapid and sensitive responsiveness on a molecular level.

A muscle cell is also known as a muscle fiber. It is an elongated cell containing alternating, regularly spaced thin filaments of the protein actin, and thin filaments of myosin, which are bonded together with smaller protein molecules. Under the microscope, the cell appears striped or striated, with obvious junctions where sets of filaments meet (these are called sarcomeres). When the muscle cell contracts, the bonds between the two filament types break, allowing the filaments to slide over one another. The bonds then re-form in the new, contracted position, with the sarcomere now shorter. With all the cells in a muscle contracting together, the muscle itself contracts, producing movement.

Soft-bodied invertebrates, such as nematode worms and cephalopods, have pliant muscular bodies whose shape is only limited by the flexibility of their outer epithelia. Animals with skeletons or exoskeletons have more limited movement, based around joints, around which the rigid body sections can move to some degree, under the control of muscles attached by ligaments and tendons to different points on the bones or exoskeleton. In a human arm, for example, contraction of the bicep muscle (attached to the shoulder blade and the forearm bones) bends the arm at the elbow, and contraction of the triceps straightens it. In insects and other arthropods, minuscule muscles allow multidirectional movement of many-jointed appendages such as antennae, as well as the legs.

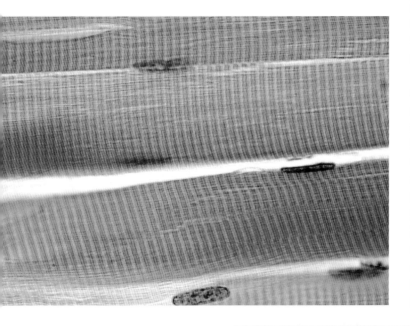

Above *Muscle cells, showing the sarcomeres where actin and myosin fibres meet.*

Right *A spider preparing to begin "ballooning"—an example of long-distance animal movement which is not muscle-powered.*

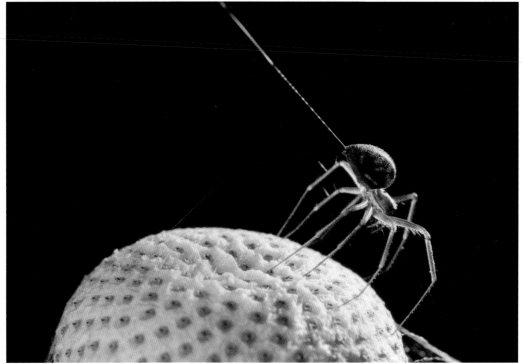

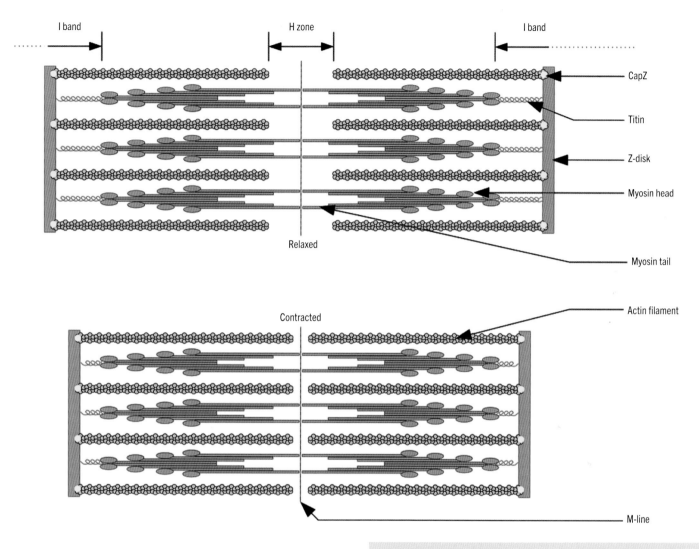

I band　　　　　　　　　H zone　　　　　　　　I band

CapZ

Titin

Z-disk

Myosin head

Relaxed

Myosin tail

Actin filament

Contracted

M-line

Above *How muscle cells contract, through the sliding movement of actin and myosin filaments.*

Tiny newly hatched spiders need to move away from their parental web. They can make extremely long journeys by "ballooning"—they extend silk threads into the air, which catch the breeze and float them away. They thus temporarily join the population of "aerial plankton"—microscopic living things in the air, many of them swept up in dust storms or from sea spray. Their movement is passive, like that of water-dwelling plankton, and gives them a means to disperse to new habitats.

JETS AND SAILS

Taking water into a body cavity and squirting it out can create rapid bursts of movement. Squids and octopuses have a large body cavity that they use in this way. The jellyfish-like organism *Nanomia* has multiple jets and uses them in different ways to maximize speed or efficiency as necessary. Like other members of the group Siphonophorae, a single *Nanomia* is not one animal as such, but a colony of tiny independent units or zooids, which co-ordinate their movements and other processes to function like a single organism. Another member of this group, *Velella*, has an unusual and very efficient passive means of movement—a rigid, thin fold of its body sits above the water line and catches the wind, allowing it to sail (sometimes over surprisingly long distances).

// Structure of mammalian skin and hair

As we have seen, the majority of animals live in water. Living on land presents animals with several challenges. The animal cell is permeable to water by its nature, but land animals need to keep hold of that water so as not to dehydrate. They must also cope with much more rapid temperature changes than water animals experience, as well as a range of different weather conditions. Some of these problems can be mitigated by habitat choice (which is why you generally find amphibians in waterways or damp places, and both reptiles and amphibians are much more abundant in regions with stable climates).

Below *Curly hair has arisen as a mutation in many domestic mammals but is scarce in wild species. This young bison is one of the few with natural curls.*

Mammals, though, are able to live under a much wider range of conditions, and part of the reason for this is the structure of their skin. The skin is said to be the largest organ in the human body, and it is right to think of it as an organ rather than a simple covering, as it is a multifunctional structure comprised of several distinct tissue types and many different cell types.

Right *A human hair under the microscope.*

Below *A cross-section through human skin.*

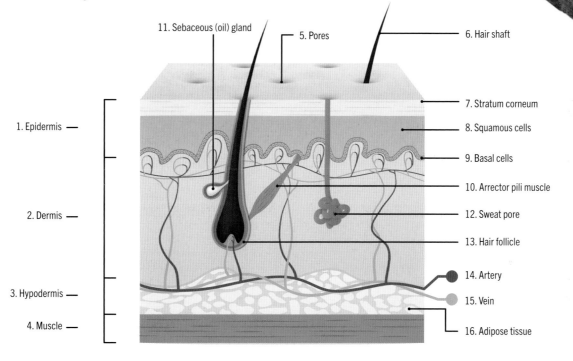

1. Epidermis
2. Dermis
3. Hypodermis
4. Muscle

11. Sebaceous (oil) gland
5. Pores
6. Hair shaft
7. Stratum corneum
8. Squamous cells
9. Basal cells
10. Arrector pili muscle
12. Sweat pore
13. Hair follicle
14. Artery
15. Vein
16. Adipose tissue

The two main layers of human skin are the epidermis, and below that the much thicker dermis, with the hypodermis below that. The epidermis comprises keratinocyte cells, which generate keratin protein that forms a matrix on the outermost skin surface. Keratin, along with older, dead, and toughened keratinocytes, provides a barrier that protects you from injury—it constantly wears away and is replenished from the active layer of keratinocytes below. Deeper in the dermis, there is a rich blood supply where structures including sweat glands and hair follicles originate, and penetrate through the epidermis to the outside world. Each hair follicle has an associated miniature muscle, allowing the hair to stand up or flatten, and a sebaceous gland, which secretes a lubricating oil. Below the dermis is a layer of adipose tissue, which provides insulation. This layer is extremely thick in seals and cetaceans (whales and dolphins) to allow them to swim and dive in deep, cold water.

The visible part of a single mammalian hair is made of keratinized, non-living cells. As it grows, pigment is deposited in the center of the hair shaft (medulla), and this gives it its color. Mammal hair contains only melanin pigments, so its color palette is limited to black, various shades of brown and grey, and golden-brown or reddish tones. Individual hairs may have bands of different melanin deposition—these multi-toned hairs produce the slightly shimmering-looking coat color known as agouti, which provides better camouflage than a block of solid color would. Hairs that lack pigment are white in appearance—the depigmentation of our hair as we age is down to a decline in the number of melanocytes (melanin producing cells) in the lower epidermis. Hair that curls as it grows, which occurs naturally in some mammals including humans, has a different follicle shape to the follicles that produce straight hairs, and curly hairs in cross-section are elliptical in shape, rather than circular.

// Specialized animal organs

In this chapter, we have explored animal cells, tissues, structures, and systems primarily through examining how they look and function in our own bodies. Some are near-universal (albeit differing in some details) across the animal kingdom, but animals are enormously varied in their overall bodily appearance and way of life, and some groups have evolved certain highly specialized structures that are unique to their own lineage.

Animal eyes have arisen in many different ways in different groups. An eye is, fundamentally, a structure that senses light, so could even include the eyespots present in single-celled eukaryotes such as *Euglena*. In animals, some of the most impressive eyes belong to mantis shrimps. These animals' eyes have 16 different kinds of photoreceptors for perceiving different light wavelengths (compared to the measly three types of color-sensing cone cells in a human retina), including ultraviolet and polarized light. Like insect eyes, mantis shrimp eyes are compound, formed of 10,000 individual light-sensing units, and can move independently to provide an even more detailed picture of their world. In terms of all-round vision, the winner is the brittle-star *Ophiocoma wendtii,* which has light-sensing structures over its entire five-armed body.

Starfish are also notable for their stomachs—of which they have two each. One of these, the cardiac stomach, can be completely everted outside of the animal's body, to engulf a food item, or even force its stomach in between the two shells of a bivalve mollusc. The stomach then secretes enzymes and absorbs the resultant soupy digested material through its epithelial cells.

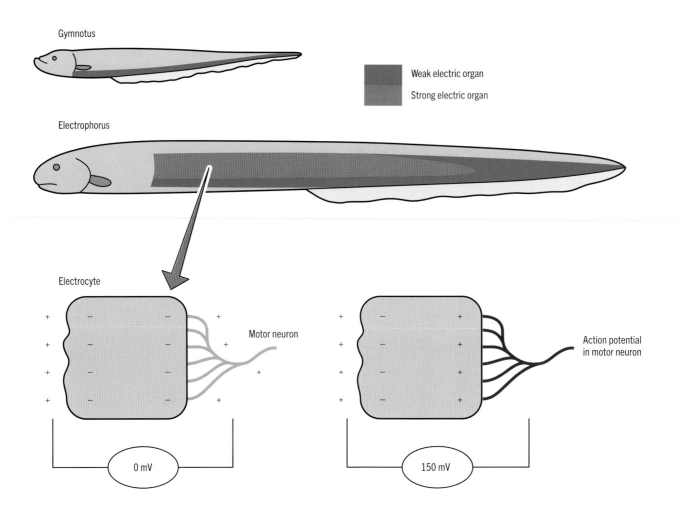

Above *Electric eels of the genera* Gymnotus *and* Electrophorus *possess specialized organs containing electrocyte cells. They generate their voltage through flow of positively and negatively charged ions through the cells' membranes.*

Many animals are able to produce toxic or venomous substances, which they use in self-defense and (in the case of venom) to immobilize prey. The venom glands of snakes are modified parotid glands (part of the salivary gland system), linked up to hollow fangs through which the venom can be forcefully expelled. Cone shells inject their venom via a modified tooth on their tongue-like radula. Venom molecules are protein-based and the recipe for their synthesis is coded in the genes. In the case of toxins, some of these are not produced in the animal's body but come from the plants that it eats and are sequestered (stored in their original form) in specialized organs. Animals that sequester toxins often show memorably bright warning coloration, so a predator that eats one of them will learn not to make the same mistake again.

Above *This cinnabar caterpillar stores toxins from its food plant (ragwort) in specialized tissues. Its bold coloration warns predators of its toxicity.*

Left *The mantis shrimp's curious eyes contain 16 different types of light-sensing cells, giving it a visual experience that we can only try to imagine.*

The ability to sense electrical fields has evolved in many animal lineages, from birds to platypuses and sharks to bees. They use this ability to sense the electric fields produced by other living things. Then there are various unrelated fish that have evolved a way to generate an exceptionally powerful electric field of their own at will, to deliver a shock and stun their prey. They possess specialized electric organs, which vary in location. The electric eel's three pairs of large electric organs occupy most of its interior body space and, collectively the electrocyte cells they contain can generate a 600-volt shock.

Glossary

Aerobic Describes a reaction that requires oxygen (e.g. aerobic respiration)

Archaea A domain of prokaryote life (and the members of that domain).

Amino acids The molecules from which proteins are built.

Anabolism Building complex molecules from simpler components

Anaerobic Describes a reaction that does not require oxygen (e.g. anaerobic respiration)

Antigen A molecule that 'fits' to an antibody, allowing it to recognise a pathogen

Autotrophs Organisms that make their own food, e.g. plants

Bacillus A rod-shaped bacterium

Bacteria A domain of prokaryote life (and the members of that domain).

Bacteriophages Viruses that attack bacteria

Blastocyst The ball of undifferentiated cells formed after a fertilised egg has divided a few times.

Catabolism Breaking down complex molecules into simpler components

Cell Smallest single unit of an organism. Some organisms are unicellular, some multicellular.

Chlorophyll The light-sensitive pigment found in photosynthesizing organisms.

Chloroplast An organelle found in plants and algae, where photosynthesis occurs.

Chromatids Each half of a pair of chromosomes

Chromoplast A type of plastid that holds igment (includes chloroplasts)

Chromosome A length of DNA, containing multiple genes — typically comes as a pair.

Cilia Fine hairs found on the outside of some organisms. Used for propulsion, among other things.

Convergent evolution When unrelated organisms evolve similarities because of adapting to similar ways of life

Cyanobacteria A group of Bacteria that can photosynthesize. The ancestors of chloroplasts.

Cytoplasm The material that fills living cells

Diploid A cell with two sets of chromosomes

DNA A molecule which carries coding for building proteins. The constituent molecule of genes and chromosomes.

Domain The fundamental division of living things in most classification systems.

Endoplasmic reticulum A structure in eukaryotic cells, with various metabolic functions.

Endosymbiosis When one organism lives and functions inside another one.

Eukaryote Organism with complex cells containing nuclei and other organelles, comprises the domain Eukaryota.

Extremophiles Organisms that survive in very hostile environmental conditions.

Flagellum A whip-like propulsive 'tail' found on some single-celled organisms.

Gamete A sex cell (egg or sperm)

Genetic mutation A change to a genome, due to a copying error during DNA replication.

Genome The complete genetic code of an organism.

Genus A group of closely related species (plural genera).

Haploid A cell with only one set of chromosomes (typically, a gamete).

Hyperparasite A parasite whose host is itself a parasite.

Kingdom A taxonomic category, one step below Domain.

Lysosome An organelle containing digestive enzymes.

Macrophage A cell that functions within the immune system, by engulfing and destroying pathogens.

Meiosis Cell division that results in haploid daughter cells.

Metabolism The chemical processes involved with the breakdown, storage and use of nutrients in the body.

Mitochondrion A type of organelle, site of respiration and descended from free-living members of Bacteria

Mitosis Cell division that results in diploid daughter cells.

Monophyletic Describes a complete taxonomic group, containing all the descendents of a single common ancestor.

Mycelium Multiple fungal hyphae.

Myelin sheath The coating around the axon of a neuron.

Nucleus An organelle in a eukaryotic cell, where chromosomes reside.

Oomycetes A group of fungus-like single-celled organisms.

Organelle A discrete structure within a cell, with a particular function or set of functions.

Osmosis The tendency of water molecules to move from weak concentrations to strong concentrations.

Parasite An organism that exploits another for its survival.

Pathogen An organism that causes disease in another.

Phagocytosis Engulfment of objects (including food and pathogens) by infolding of the cell membrane.

Phagosome A membrane-bound, engulfed item within a cell.

Photosynthesis The chemical process of building glucose molecules using sunlight.

Phylum A taxonomic grouping, one step below Kingdom

Phylogenetic Relating to the evolutionary relationships between organisms.

Phytoplankton The photosynthesising component of free-swimming microscopic water life.

Pistil The female reproductive structure in a flower.

Plasmid A small ring of DNA found in some prokaryote cells, and distinct from the cell's main chromosome.

Plastid An organelle in plant cells, derived from cyanobacteria. Includes chloroplasts.

Polymer A molecule built from repeating units arranged in a chain (for example, starch is a polymer of glucose).

Prion A protein molecule which acts as a disease-causing agent.

Producer An organism that can build its own food supply (e.g. through photosynthesis).

Prokaryote Simple organism whose cells lack nuclei, comprises the domains Bacteria and Archaea.

Protozoa Umbrella term for many 'animal-like' single-celled organisms.

Pseudopodia Projections in the cell membrane of some single-celled eukaryotes, used for movement.

Ribosome A cell organelle, responsible for protein synthesis.

RNA A single-stranded, self-replicating nucleic acid.

Saprophyte An organism that breaks down dead tissue.

Sarcomere A band across a muscle cell, where the two protein filament types meet and contraction occurs.

Septa A boundary or wall between two parts of a structure.

Subviral agent A biologically active entity which is simpler than a virus (e.g. a viroid or prion).

Symbiosis A stable relationship between two species which benefits one or both.

Virion A single viral particle.

Viroid A simple virus-like entity.

Zooid The individual units that form a colonial organism such as a hydrozoan.

Zooplankton The non-photosynthesising component of free-swimming microscopic water life.

Index

Picture credits